Theory of
Irregular War

Theory of
Irregular War

JONATHAN W. HACKETT

McFarland & Company, Inc., Publishers
Jefferson, North Carolina

This book has undergone peer review.

The Defense Office of Pre-Publication and Security Review cleared this work for publication. The views expressed in this publication are those of the author and do not necessarily reflect the official policy or position of the Department of Defense or the U.S. government.

Some names were modified or withheld to protect ongoing operations and the personal security of those involved in sensitive activities.

ISBN (print) 978-1-4766-8905-0
ISBN (ebook) 978-1-4766-5154-5

Library of Congress and British Library
cataloguing data are available

Library of Congress Control Number 2023054550

© 2024 Jonathan W. Hackett. All rights reserved

No part of this book may be reproduced or transmitted in any form or by any means, electronic or mechanical, including photocopying or recording, or by any information storage and retrieval system, without permission in writing from the publisher.

Front cover image: Thousands of mortar shells fired from inside an abandoned university that changed hands numerous times between two insurgent groups during the Lebanese civil war (taken by the author during his deployment in Lebanon)

Printed in the United States of America

*McFarland & Company, Inc., Publishers
Box 611, Jefferson, North Carolina 28640
www.mcfarlandpub.com*

For Mojgan
عشقم

E le genti che passeranno,
O bella ciao, bella ciao, bella ciao ciao ciao,
E le genti che passeranno
Mi diranno "che bel fior."

Questo e il fiore del partigiano,
O bella ciao, bella ciao, bella ciao ciao ciao,
Questo e il fiore del partigiano
Morto per la liberta.
—Italian folk song

Table of Contents

Acknowledgments	ix
Preface	1

Part I: Thinking About Irregular War

1. Enemies of the State	5
EXISTING THEORETICAL FRAMEWORKS	6
A NOTE ON METHOD	13
A NOTE ON SOURCES	16
STRUCTURE OF THE BOOK	18
2. A Framework of Thought, Action, and Justice	21
A SCHOOL OF THOUGHT AND ACTION	22
JOHN RAWLS AND THE THEORY OF JUSTICE	25
THE FUNCTIONAL SOVEREIGN	28
THE DYSFUNCTIONAL SOVEREIGN	31
VIOLENT RESISTANCE AND THE DYSFUNCTIONAL SOVEREIGN	33
'ASABIYYAH AND ANOMIE	35
3. The Structure of Irregular War	38
THEORIES AND PHILOSOPHIES OF WAR, OR "ON CLAUSEWITZ"	38
IRREGULAR WAR: A SOCIAL, TERRITORIAL, AND POLITICAL AFFAIR	44
IRREGULAR WAR AND SOCIAL ORDER	45
IRREGULAR WAR AND SOVEREIGN TERRITORY	48
IRREGULAR WAR AND POLITICAL-ECONOMIC INSTITUTIONS	50
THE CONDUCT OF IRREGULAR WAR	52
STRUCTURES OF CONFLICT: INSURGENCY AND REVOLUTION	54

Part II: The Synecdoche Trap

4. Problems of Scope, Method, Bias, and Character	61
PROBLEM ONE: SCOPE	62

Table of Contents

    Problem Two: "Methodismus" — 64
    Problem Three: Bias — 67
    Problem Four: Character — 73
    The Human Element — 76

5. Conventional War Theory and Regular Wars — 78
    Law and Doctrine — 79
    Doctrinal Disconnect — 81
    Wars, Conventional and Regular — 84

6. Conventional Forces in Irregular Wars — 86
    Combatting Bandits in Small Wars — 87
    A French Vision of Disaster — 98
    The British Method — 102
    American Force — 104
    Suppression and Destruction — 110
    Third-Party Intervention — 111

Part III: A Theory of Irregular War

7. Irregular War Conditions — 127
    Conditions of Social Order and Sovereign Dysfunction — 127
    Conditions of Sovereign Territory and Sovereign Dysfunction — 134
    Conditions of Political-Economic Institutions and Sovereign Dysfunction — 143

8. The Elements: People, Politics, and Propaganda — 150
    Thinkers and Actors — 150
    The Why — 155
    Wings of Resistance — 162
    Ideology and the Message — 166

9. A Dialectic of Irregular War — 173
    The Intersection of People and the State — 173
    The State and Its Enemies — 178
    Irregular Wars and Sovereign Dysfunction — 186

Concluding Irregular Wars — 197
Chapter Notes — 201
References — 227
Index — 243

Acknowledgments

My days writing on irregular wars were not set in the quiet, air-conditioned offices and libraries many authors on the subject have enjoyed. I am neither a professor nor a diplomat nor an officer. Instead, my journey has taken place far from those relative comforts but closer to humanity. I was not alone on this journey, and for that I am grateful.

I would not have come this far if not for my wife, Mojgan, whom I met in Iraq while I was on the ground controlling lethal strikes into Mosul and its environs against Da'esh. Her patience and steadfastness since that time are immeasurable.

The brave Syrians fighting in Dara'a and across the country taught me so much about the price of resistance and the value of freedom. I am indebted to Ma'an, Issam, Ahmed, and others fighting to resist an oppressive government. It was in my talks with these Syrians and others that I came to formulate the concept of sovereign dysfunction.

While preparing for a mission in Southeast Asia, I was fortunate to come under the mentorship of Dan, an unmatched expert on Indonesia and the region. I am indebted to him for imparting his nuanced understanding of the unique social orders coexisting in an intricate web across the 13,000 islands comprising what is now Indonesia. Dan taught me to learn by *jalan-jalan*, walking in the places I wished to understand rather than studying them from afar.

My understanding of the limitations of territorial sovereignty matured while operating in the mountains and deserts of Northwest Africa. My dear friend Toufiq brought me to his family home in Beni-Mallal in the beautiful mountains of the Atlas range, a place the Phoenicians, Romans, and Vandals could not penetrate, and the Muslim conquerors could only claim on their maps. The residue of French rule remained only in officialdom and in the more than occasionally frustrating civil infrastructure designs. But Toufiq and other Berbers had not surrendered their identity. While we gathered almonds on the rough cliff faces, I came to appreciate the power terrain wields over states and societies.

Acknowledgments

The many Marines, Sailors, Soldiers, and Airmen I have had the honor to serve with have each taught me something. Marine Raiders such as J.B., L.E., and G.G. each left me with something to remember. Among the most impactful experiences was my time with a Marine Special Operations Team in Lebanon where I had the opportunity to work with some of Lebanon's finest in the Maghaweer, the Moukafaha, and the Intervention Regiments. It was in moments such as picking thyme on bombed-out hills of Kfarfalous with salty old Sergeant Hawk that I began to understand the implications of a dysfunctional sovereign interacting with political and economic institutions. I am also grateful for my conversations and experiences in Lebanon with people like a Shi'a military leader, a Christian arms dealer, and a man from Northern Ireland who fought against the South Lebanon Army in the 1980s.

I also want to acknowledge those I knew who lost their lives due to involvement in irregular wars. Some, including Charles Keating IV and Eric Christian, never made it out. Others, such as Bobby Cox and 15 others who died in a KC-130 crash and the Marines of the Raider 7 Blackhawk crash, were on their way to another one. Marine pilot Taj Sareen was on his way home. Some, like Rick Villani, Jastin Pak, and Mike Easley, could not escape the wars they fought within.

Readers of early drafts offered important advice I could not do without, including Michael Manzo, Phil Messer, and several intelligence professionals who wished that their identities be withheld. Any errors that survived into the current text are my own.

Preface

"You have the watches; we have the time." This Afghan proverb captures something of the essence of violent conflict between forces struggling for sovereignty, a conflict in which at least one group seeking that sovereignty is not a state. Something many states engaged in such conflicts miss is that the act of fighting is *warfare*, not *war*. There are many ways to fight a war, and many more ways to lose. Whether those ways are conventional/unconventional, regular/irregular, symmetric/asymmetric, overt/covert, Napoleonic/Fabian, or any other diametric word pairing is irrelevant at the level of analysis dealing with war itself. War is a sovereign weapon expressed through organized violence between parties clashing over incommensurable policies.

States often find themselves embroiled in violence. Scholars, practitioners, policymakers, and voters alike prefer tangible effects, visible results, something to be measured as the proper weight of vengeance, all at the expense of rational objectivity and a look toward possible futures for our global community. In doing so, the people are reified as "the enemy," while sovereignty demanded is forcibly alienated from the sovereignty imposed. This creates a contradiction, but not one of an intangible, esoteric, Marxian kind. Instead, it is the contradiction between the way a people expect their sovereign to function and the antithesis of this expectation: the dysfunctional sovereign. This contradiction is very tangible, very clear, and very violent. This is the problem of a dysfunctional sovereign interacting in unexpected ways with social orders, sovereign territory, and political-economic institutions. It is the problem forming the basis of this book, the problem of sovereign dysfunction and irregular war.

Many soldiers have fought and died in wars they did not understand, in lands their countrymen never heard of, against an enemy they never saw. There is something fundamentally different about these wars when viewed against those in which states employ their military instrument of national power against other states. In this book I argue that this difference has nothing to do with how the war is fought (i.e., *warfare*), and everything to do with why it is fought and what caused it (i.e., *war*). We must understand the *why* behind the *what* before we can plan the *how*. But before attempting to do so, allow me to share a brief story to illustrate the dynamics of the conflicts I am about to explain.

My first deployment as an interrogator was to Helmand Province, Afghanistan. Many I met there, regardless of their ethnic background or tribal affiliation, agreed on one point: ever since Alexander the Great failed to conquer Afghanistan 2,400 years ago, no foreign power had succeeded in the same endeavor. "Not even the United States!" they would proudly add. This common refrain reminded me often of the unifying phenomenon of resistance through proximity, especially resistance to a power more alien than the force resisting it.

Still, I never really understood the kind of war I was part of until interrogating one particular Taliban detainee. Before talking with that detainee, I had to finish my reports from an eight-hour interrogation of another detainee, an indentured opium harvester whose family had survived the harsh winter in southern Afghanistan on a diet of poppy seeds, and only poppy seeds, for about six months. That indentured laborer never learned his six daughters' names, he loved his wife, and he lamented the fact that half of his eleven children would die before adulthood due to war, starvation, and generally horrible living conditions. "That's why I had eleven children, actually, because I could only support a handful of them and I knew half would die," he told me as he gingerly held a cup of warm tea I had given him.

After finishing my reports from talking with the indentured laborer, I readied myself and got back into the interrogation booth, this time with the well-fed, well-read, and defiant local Taliban official with a beard as carefully shaped as his words. He sat cross-legged on the floor in the dark blue jumpsuit we had every detainee wear, but his posture and affect were that of a judge behind the bench rather than an "unlawful enemy combatant," as the U.S. government categorized Taliban detainees like him. After some time, his defiance broke down somewhat and he became receptive to a little *tête-à-tête* on politics.

"What is the definition of insurgency?" he asked me through my interpreter. I replied with my best military doctrinal answer that it was the violent overthrow of a sovereign state by a group that is not a state. The Taliban came to my mind.

"Who violently overthrew the sovereign government of Afghanistan in October 2001?" he asked.

"Well," I began, "it was the Northern Alliance, supported by the United States."

He sat back, apparently pleased with my answer. "Then, are you not the insurgents?"

Part I

Thinking About Irregular War

Chapter 1

Enemies of the State

> Every revolution creates a temporary power vacuum when the power of the *ancien régime* has been destroyed but the revolutionary regime has not yet asserted itself throughout the territory.—James C. Scott[1]

The idea of the state as an expression of ultimate sovereignty has dominated international politics and domestic power structures for the last four hundred years. Since the Peace of Westphalia ended the Thirty Years War in 1648, states have invested immeasurable resources into developing, equipping, and employing conventional forces designed to understand, deter, confront, destroy, and rebuild the conventional forces of other states. This is true despite most cases of organized political violence over the past four centuries occurring not between states, but within them. Those conflicts within states are neither a spontaneous combustion of grievances nor are they ends in themselves: they are ripples on the surface of the deep.

The *state* in this sense is the combination of a nation and a community in the form of a polity, hence a *national state*, and theories explaining wars between states represent conventional war theories.[2] Indeed, studies of the nature of conventional military conflicts are among the most popular works in modern theoretical approaches in sociology.[3] Theoretical and conceptual studies of irregular war pale in comparison to that of states, the wars between them, and the international political system. Few studies approach irregular wars objectively, and military theorists are among the worst offenders. Such works usually begin with an assumption of war *a priori*, then develop or recommend strategies treating the cause of the conflict in passing, if at all, and often favor whatever state the theorist or strategist supports or hails from. This book will accomplish two things: I will show that existing theoretical frameworks are insufficient to understand and explain irregular wars and I will present a novel theory that can. To do so, let us begin with a technical definition of irregular war. *Irregular war is the apotheosis of conflict between the people and the state, a violent dialectic between a faction and a sovereign expressed outside existing political institutions.*

In the following chapters, each element of this definition will be broken down and examined in detail to reveal the conditions at work in these conflicts. I employ a systemic approach—from the outside-in rather than inside-out—as a starting point for assessing conditions and case-specific environmental factors influencing causes, conduct, and outcomes of irregular wars. Systemic analysis is useful because it can explain, for example, why certain counterinsurgency operations that appeared to be successful on the surface were instead highly contingent upon exogenous factors outside the control or even the knowledge of the participants rather than from employment of a supposedly brilliant strategy in places like Malaya or the Philippines. The systemic approach carries similar explanatory power for analyses of unconventional warfare operations like the Chindits in Burma or the Mujahideen in Afghanistan.

It is important to distinguish here between irregular war, the conflict itself, and unconventional warfare, an operation pursued during the conduct of irregular warfare. We will dive into greater detail on the difference between irregular war, irregular warfare, and unconventional warfare in subsequent chapters, but for now it is enough to distinguish between irregular war as the conflict in general and unconventional warfare as a potential operational activity in such a conflict. Unconventional warfare does not need to be conducted solely in irregular wars; indeed, unconventional warfare is usually a tactic employed in conventional wars as a spoiler against a conventional enemy. This book is not about unconventional warfare. This book is about irregular wars, the people who may find themselves engaged in such conflicts, and how they got there. It is important to make this distinction, because the U.S. military consistently struggles to deploy these terms into useful theoretical frameworks for understanding irregular wars.[4]

Existing Theoretical Frameworks

Who fights, who leads, and why do they engage in irregular wars? There are several important theoretical frameworks in the field of sociology that partially address these questions. Most agree that there are three major theoretical frameworks for studying revolutionary movements from the perspective of the state, but even the number of frameworks is debated among scholars.[5] In contrast to the three most often used frameworks, scholars focusing on the people rather than the state accept six theories suited for approaching war studies.[6] Two common schools, Modernization theory and Marxism, both hold that revolutions occur in societies transitioning from traditional to modern forms. This is an unhelpful formula, however, since it does not explain all revolutions and most rough political transitions are not actually revolutionary.[7]

Governments and their militaries have frequently failed to understand the key variables involving organized political violence against the state. A cursory comparison of cases indicates there is something at work in this violent process pushing the conflict in predictable directions. The Iranian revolution in 1979 succeeded in transforming the social order to topple the Shah, but Kurdish rebels sustained their Mahabad Republic in northwest Iran for only a year in 1946 after Russia withdrew. The Sudan People's Liberation Army succeeded in creating an internationally recognized state in the sovereign territory of South Sudan, while the Sahrawi Arab Democratic Republic claims sovereignty over an unrecognized state in southern Morocco from its base in a refugee camp in Algeria. Magsaysay successfully worked through political-economic institutions to neutralize the Huks in the Philippines after the Second World War, yet the Dutch lost their colony in Indonesia to a small band of Javanese separatists in the same period after four centuries of extreme colonial domination and economic exploitation. Here we begin to see that something is going on in inside the categories of social order, sovereign territory, and political-economic institutions. Analysis of these and other cases hints at interesting peculiarities in cause, conduct, and outcome for each manifestation of organized political violence against the state.

Those unique characteristics closely follow the challenges to social orders, sovereign territories, and political-economic institutions that can only be explained by an intervening factor enabling the emergence of a serious competitor to the incumbent government. Meanwhile, the absence of that factor leaves the incumbent regime intact regardless of its inefficiency, brutality, or ineptitude. This intervening factor is *sovereign dysfunction*, and its presence in the conflict leads to irregular wars in which insurgents, conventional forces, and third-party interventions seek to preserve, destroy, or transform the state through organized political violence. States fail to understand these wars and militaries fail to win them.

Trends in military theory favor a teleological approach to war that is unscientific in method and weak in explanatory power, especially those treating irregular war. A new theory is therefore needed to explain how states with weak political institutions and low levels of centralized authority are prone to irregular wars. In those cases, the traditional methods for reconciling popular grievances, conducting power transitions, and granting social concessions are processed in a liminal space outside state control, potentially leading to a condition of sovereign dysfunction in which irregular war may occur. The conditions and variables at work in irregular wars can be explained using the conceptual framework outlined in this book to analyze these conflicts.

Several chapters of this book are devoted to first identifying and then dismantling major elements of irregular war theory especially in the U.S. military that are long-standing and nonetheless weakly constructed. Charles Darwin's advice is applicable here: "To kill an error is as good a service as, and sometimes even better than, the establishing of a new truth or fact."[8] This book both kills the error of conventional military theories applied in irregular wars and establishes a new way of understanding the facts at work in those conflicts. Honest critiques of existing theories are necessary to help synthesize an approach that is broadly applicable, testable, and useful for a wide range of cases. My critiques of previous ideas, especially those ideas deeply embedded in current U.S. military doctrine, should not be taken as attacks. Challenges to long-standing claims are an effective way to advance understanding. At the same time, sweeping, unfalsifiable claims about war are not scientific at all and should be excluded from discussions at the level of theory. This book offers a balanced point of view, holding insurgents on an equal plane with the state, avoiding the first flaw in many studies attempting to address irregular wars: favoring the state whatever its characteristics over the insurgency regardless of their cause. The state-centered approach is sometimes useful because it examines the states subject to overthrow in irregular wars in addition to the non-state groups arrayed against them. Despite the apparent utility of this approach, disagreement exists in important ways in nearly every aspect across the disciplines concerned.[9] This disagreement indicates that a suitable framework remains elusive.

Nevertheless, there are several prominent approaches that have become popular over the past two centuries, and those approaches tend to dominate the literature on conflicts between man and the state, for better or worse. The first category could be called "people-centric," which approaches social revolutions, violent regime transitions, and civil wars through their themes of social order through established social science research methods.[10] The second approach can be described as "state-centric," treating wars in the context of state-based levels of analysis, in which states make war against other states and any other organized violence is considered subordinate to this activity.[11] The third approach can be described as "military-centric" and is often found in the security or military studies fields, approaching war as a subordinate instrument of state policy.[12] Each of these approaches contains numerous arguments and counterarguments, often making their cases within an artificial context narrowly defining war as a preserve of the state. The people-centric category tends to favor framing insurgencies and revolutions in terms of tensions between the haves and have-nots, albeit in subtler terms, such as the thick description in Theda Skocpol's important study of revolutions in France, China, and Russia.[13] Skocpol's analysis of these revolutions is routinely cited in the social sciences, often because of its detailed narrative of the conditions leading up to these major revolutionary wars. Still, Skocpol's study focuses on agrarian and peasant uprisings against large states with vast systems of inequality, but she does not examine the myriad smaller conflicts, both in scope and time, which cast their shadows over the previous four centuries.

Despite the great impact the French, Russian, and Chinese cases have had on the international political system, they are exceptions rather than the rule. This is especially true as social revolutions have become less frequent in the years following the Second World War. In contrast to the structural analysis employed in the mid-nineteenth century, more recent studies of upheavals in Iran, Nicaragua, and the Philippines more closely examined the variables involved over time in each case.[14] Like these more recent treatments, I avoid the Eurocentric worldview and broad structural analysis of only the most prominent events. Instead, I address cases across cultures and reach outside the relatively rare instances of peasant revolts against empires.

Works in the state-centric category tend to follow the three "images" Kenneth Waltz developed to analyze the actions of people, states as units, and wars among them in an international political context.[15] Waltz continued developing this framework as he came to view states as agentic black boxes competing in an anarchic world.[16] The "images" framework of states that Waltz crafted continues shaping how scholars write about war, though it is strictly state-centered. This school of thought features prominently in studies of revolutions and wars, two phenomena such studies place firmly within the Realist world of states as black boxes engaging one another in the international political system.[17] Many political Realists follow the model of international politics popularized during the Cold War, which places the state at the center of the political action.[18]

State-centric models are helpful but incomplete. These often do not address irregular wars on their own, nor do they descend deeper into a level of analysis which examines the individual irregular fighter. A complete analysis must include individual motivations and the effects of certain conditions within the state, rather than between them, to explain why citizens become insurgents since irregular war conditions occur outside this simple triptych of people, sovereigns, and state violence. The "images" model of the international political system, though useful and still applicable for conventional wars, must either be modified or discarded altogether when analyzing irregular wars.

Still within the state-centered approach, those focusing on political economy rather than the state in the abstract have found broader explanations of conflict between states and their people. Analyzing the way states transition into or out of democracy from a political-economic perspective helps to elucidate the role institutions play in violent political transitions.[19] These transitions may follow certain forms of structural development depending on the political and economic institutions involved in those transitions. This approach is based firmly in political economy, allowing for some fruitful conclusions based on a combination of quantitative analysis and descriptive inference. However, a narrow focus on the economic aspects of social orders interacting with the structure of the state limits the analysis primarily to non-violent political contexts of transition.[20]

The Cold War and its aftermath produced a rich tradition of studies of revolution and war as social phenomena in the context of the state. It was during this time that examinations of the character of social revolutions began gaining traction. Although their findings reside in the context of the Cold War, there is some still-relevant information in these studies regarding the development of Westphalian politics.[21] The most useful of these works primarily examined power and violence wielded by the state against its own people through cases of social revolutions that have already occurred. While this is a useful approach, these studies are understandably bogged down in the milieu of Marxist-Leninist revolutions, which makes sense since these conflicts dominated the world in which those authors lived. In contrast, I take a deliberately broad view, both temporally and culturally, to analyze a diverse set of cases that are not often compared.

Unlike the majority of studies taking a state-centered view, a minority have approached the issue from the perspective of the insurgent.[22] One such study of insurgent movements against states from 1945 to 1991 thoughtfully invoked Leon Trotsky's familiar dictum from 1930 that "a revolution takes place only when there is no other way out."[23] This brings forth the escape valve concept, which numerous authoritarian states have exploited to reduce social unrest, whether through state-controlled press, sham elections, or rallying against enemies of the nation. Insurgency and revolution are last resorts even in the most ideologically polarized conditions. The most telling studies take a quite narrow approach, for example focusing on the civil war in the Greek Argolis or the Algerian civil war in the 1990s.[24] Such studies reveal useful implications, not least of which is the possibility that violence in civil wars between anonymous actors is not necessarily greater than between people who know each other, contrary to traditional claims in Cold War period studies. In fact, evidence suggests that violence in civil wars might actually be greater between those who know each other than those who do not. Such findings on the intimacy of violence in civil wars are especially useful for understanding why people decide to engage in irregular wars.

The military studies category dutifully follows the Prussian officer Carl von Clausewitz, despite his own admission that what he called "people's war" did not fit his own paradigm of armed conflict in his unfinished 1832 work, *Vom Krieg [On War]*. For example, following Clausewitz, one influential work framed combatants as soldierly instruments of the Westphalian state.[25] This wholly conventional perspective failed to approach combatants who are not members of state military institutions, especially those fighting to overthrow, secede from, or transform an existing state. Further, Clausewitz noted toward the end of his treatise that his conception of war applied only to conventional conflicts between Westphalian states, employing regular military forces, and conducting organized violence within the confines of state policy. He explicitly differentiated between these types of wars and the very different wars between a state and its people.

An irregular war theory need not overturn any of Clausewitz's precepts, since these apply to idealized conventional wars. Instead, any study of organized human violence could benefit from drawing upon some of Clausewitz's universal observations about the nature of violence and war that are commensurable with irregular wars. A shift in focus is required, away from the conventional *Weltanschauung* of soldiers fighting other soldiers on behalf of their respective states. Instead, it will be helpful to examine the mechanisms at work for belligerents who turn from ballot boxes to boxes of bullets.

In contrast to the conventional views of Clausewitz and his acolytes, the German general Friedrich von der Heydte wrote the first work specifically seeking to explain irregular wars in their modern context only in 1972. His book focused first on warfare and only tangentially on war, describing conduct but not explaining phenomena. Von der Heydte did make the important distinction between irregular *war* and irregular *warfare* in his book for the first time in the scholarship of war in any language. Unfortunately, the four decades between von der Heydte's publication in 1972 and the next work ostensibly dedicated to the study of irregular warfare have produced relatively little new theoretical treatments of irregular wars despite these conflicts being a virtually uninterrupted fact through the present. Far more common are reprints of old manuals and the development of what are purported to be new concepts that are, in reality, decades and even centuries old.[26]

Recent attempts to explain insurgency tend to focus on external factors, ignoring deep histories, and avoiding objectivity. Treatments in this vein are unfortunately common.[27] While some scholars at least acknowledge that these conflicts are part of something bigger, most usually fall short of the task by focusing solely on tactical considerations without stepping back from each case to ask why the insurgents may be fighting in the first place.[28] These glaring omissions neuter their arguments, as any force can employ guerrilla tactics, whether conventional or irregular. Further, irregular forces can employ conventional tactics, as the Ukrainian Insurgent Army and Tito's Partisans both did to great effect in the Second World War. David Kilcullen took a quintessentially naïve approach to irregular wars in his influential work, *The Accidental Guerrilla*, in which he claimed insurgents are generally unassuming peasants who happen to be swept up into grand conflicts they cannot comprehend all while having no real commitment to the cause in which they spontaneously find themselves embroiled.[29] Beyond the Eurocentric bias Kilcullen employed, he placed severe limitations on his research by selecting only a handful of cases all on the dependent variable of "successful counterinsurgencies," a variable he did not explicitly define. In essence, Kilcullen wrote as if to defend the U.S. Army Field Manual on counterinsurgency that he helped to write several years prior. His work is representative of, and at times parrots, many state-centered colonial policing books that bob prominently in his wake.[30]

Approaches unquestionably favoring states or militaries in irregular wars commit the error of focusing on abstract factors in a vacuum almost exclusively in their analyses of insurgencies and counterinsurgency operations to the detriment of the field.[31] An example of this error can be seen in attempts to simultaneously justify weak case selection in terms of the "critical case" of Lebanese Hezbollah while employing the same case with the opposite outcome in an analysis of state militaries in the Middle East.[32] Systemic analysis is best applied when a broad range of cases are analyzed for their data to avoid selecting on the dependent variable. In contrast to all of these, a proper study of irregular war requires a critical approach that does not shy away from asking big questions and considering the *longue durée* of history, a view integral to constructing a theory of irregular war.

These lacunae beg for a focused, comprehensive treatment of the topic to fill the yawning gap in the literature that will also prepare the field for a new theory, whether qualitative or quantitative. There are three tasks that must be accomplished to outline a qualitative theory: laying out components, explaining past phenomena, and predicting future outcomes.[33] A quantitative theory must include a formal explanation, testable hypotheses, and other elements such as lemma, corollary, postulate, and paradox.[34] In contrast, most current works on irregular wars contain descriptions and definitions rather than explanations. Definitions are logical rather than empirical: they may contain suggestions but not hypotheses.[35] To overcome the disparity between current literature and good theory, disciplines should be bridged to show that conventional wars between states do not share the same paradigm for explaining conflicts within states or between states and organized armed groups that are not states. To that end, the following chapters outline a novel approach through a broadly applicable yet simple theory explaining how irregular wars result from the relationships between dysfunctional sovereigns and their social orders, sovereign territories, and political-economic institutions.

A Note on Method

Key to the central argument I make in this book is an explanation of how the independent variable of sovereign dysfunction interacts with the dependent variables of social orders, sovereign territories, and political institutions, setting conditions conducive to irregular wars. To make this argument, I employ a combination of Grounded Theory, narrative inquiry, and comparative analysis to collect and analyze evidence for use in theory construction.[36] Cases of irregular wars are best understood through narratives and comparisons that reveal causal pathways, path dependencies, and contingent events. Comparing overtly different cases while holding certain variables constant can sometimes reveal interesting results. At the same time, every irregular war case is different because of the path dependencies involved for the participants as they make their way toward the violent moment of war.

The research program for this book used Grounded Theory to elucidate gaps in existing theories of irregular war and irregular warfare so that a new theory could be constructed.[37] To start, I categorized the dependent variables and organized the research program to explore these categories so I could argue one at a time how each of these variables are affected by the independent variable of sovereign dysfunction. The attributes of those three categories are examined as properties common across cases, while some exceptions are also examined and discussed. The steps taken for each case included sampling of available data, categorizing the data in relation to the variables, and then developing substantive theory through analysis of the interactions observed between the variables.[38] Throughout this process, I used narrative inquiry to establish temporal boundaries, identify rich context, and simulate interviews through the analysis of memoirs and other primary sources produced by those participating in the conflicts discussed.

Existing theories were analyzed using a dialectic approach. Part III of this book provides a synthesis of that dialectical process. I will show that existing theoretical frameworks are incomplete, insufficient, or inappropriate for explaining irregular war, while acknowledging that elements from some may be synthesized together. Some of my critiques of past authors and studies may appear unforgiving and dispassionate, but this is because dialectic brings us closer to the truth. If one expects a claim to be unquestioningly accepted, it is likely that those claims are neither rational nor scientific. In contrast to existing frameworks, my methods of dealing with the data collected include comparative analysis and narrative inquiry to lay the groundwork for a novel conceptual framework that helps explain the phenomenon of irregular war.

Narrative inquiry is used in certain parts of this book to discuss relevant elements within specific cases, while comparative analysis is the primary method employed to identify and analyze factors and conditions between a range of cases of irregular war. Some of these factors include organizational structures, causes of conflict, slogans and media usage, leadership characteristics, goals, performance, persistence, outcomes, and intellectual antecedents. While some studies rely on narrative inquiry to dissect single cases, this method must be applied more dynamically when examining the large range of cases with which this book engages. Some of the cases narrated throughout the work include the development of Hezbollah over the years, the dynamics of the Dutch and Javanese during the Indonesian War of Independence, and the development of several successful and unsuccessful uprisings in Iran during the twentieth century. These cases are not dealt with at once in one long narrative, as is often the case. Instead, the relevant elements of each case are teased out during the precise part of the argument in which they apply and when they fit the category identified through the application of Grounded Theory. Over the course of the book, a narrative is introduced piece by piece, but only when needed for our purposes. Similarly, the functional components of insurgency such as propaganda, ideology, and individual motivations are analyzed at the appropriate moments to focus on how these develop in different circumstances while revealing important similarities reflecting the nature of irregular war in general.

My argument relies on evidence from numerous cases to explain how the interactions of the variables promote the conditions necessary for irregular wars. To accomplish this, I draw from philosophies of justice and society developed by American philosopher John Rawls and Iranian intellectual Ali Shariati to explain the interactions between the independent and dependent variables.[39] The case elements I discuss are drawn from the "undeveloped themes" in the social sciences, in which the fertile areas of biography, leadership, cognition, zeitgeist, strategy, and agency remain understudied areas in conflict analyses.[40] I will touch on some of these topics, though only as they apply to the central idea of irregular war. This book would be incomplete without also addressing external factors, not least of which is third-party intervention. Third-party intervention is a major component of irregular war, though I will show that it is not a sufficient factor for such conflicts to begin, develop, and end. To achieve these objectives, I target all available angles of important cases while employing comparison by analogy for related cases when thicker descriptions are unavailable.

Narrative inquiry traces processes, assesses contingent events, and identifies the causal pathways shaping each outcome. Connections are drawn between various disciplines of the social sciences by narrating the elements of a case, modeling the interactions within the case, and then evaluating the potential causal factors the analysis elucidates. This ensures descriptive inferences can be drawn from each unique case.[41] This method has been used successfully in studies with narrow scope, structure, and sources focusing usually on one specific case such as analyses of demilitarization in weak and failed states in Sudan, Timor-Leste, Afghanistan, and Democratic Republic of Congo.[42] A combination of comparative analysis and narrative inquiry was used to explain, in part, why Lebanon's Hezbollah was more militarily successful than conventional Arab militaries.[43] A 2010 U.S. Army Special Operations Command study of revolutions and insurgencies from 1933 to 2009 used this method to identify five general causes of those events, though qualitative comparative analysis may have been better suited for that study.[44] Narrative inquiry, framed through Grounded Theory, is successful when applied systematically and consistently across a diverse set of cases, which I set out to do as I outline my arguments.

Whereas states such as the U.S., Israel, or Russia employ state-centered frameworks to understand and engage in irregular wars, this study presents a novel approach that considers how sovereign dysfunction interacts with social, territorial, and institutional aspects of the state leading to conditions ripe for irregular wars. A "school of thought and action" framework will lay the foundation for our later discussion of individuals, their leaders, and non-state groups such as insurgencies leading up to, during, and concluding irregular wars. A "justice as fairness" framework will assist in explaining how the real and ideal state functions in the same conflicts. It is hoped that the current study and others like it will lead to further research designed to test, challenge, and refine the claims that follow.

Research focusing on issues like regime legitimacy, methods of warfare, or a particular political system are inadequate for explaining the causes, conduct, and outcomes of irregular wars. Each case is nuanced, unique, and historically contingent, providing valuable evidence to explain why a faction may accept their fate, vote away their disagreement with the incumbent government, or resort to violence to achieve political objectives. The research for this book is structured around three conditions manifested in organized violence between a state and a group that is not a state. Specifically, the research design focuses on the factors of integration, control, and productivity of the sovereign and the link between its expected outcomes and those that actually occur. These factors are alternatively framed as the interactions between the independent variable of sovereign dysfunction and the dependent variables of social order, sovereign territory, and political institutions. Their effects are apparent across the range of cases.

A Note on Sources

A domain of inference must be set before selecting data.[45] The domain of inference for this study begins roughly with the 1648 Peace of Westphalia and continues through to the present. Much existing scholarship has focused discretely on one of three periods from 1648 to 1945, 1945 to 1991, and 1991 to the present, depending on the objectives of those studies.[46] This study seeks to avoid periodizing from any one perspective, while considering the temporal contexts logically framing each case within the *longue durée* of the Westphalian structure currently dominating the international political system. Some explanatory elements necessarily defy the constraints of time in war, and so the domain expands or contracts as necessary. The sources I chose effectively contribute to the theory construction going on in this book. This allows for a broad evaluation of the theoretical assumption that certain conditions for irregular war occur because of the interactions between a sovereign and the social, territorial, and political elements of the state.

There are numerous challenges to collecting data in irregular wars, such as the effects of time, distance, danger, and death. The phenomenon in question has a temporal arc stretching from the Peace of Westphalia in 1648 to the present, so most of the eyewitnesses to the relevant events are dead or are otherwise inaccessible. Most participants did not leave behind any evidence of what they thought, witnessed, or did. A minority left behind memoirs, essays, and other accounts that reflect their own positions on the subject rather than the whole conflict. The pages that follow draw upon a range of sources, primary and secondary, while relying on two existing social science frameworks that explain specific conditions animating the individual, the society, and the state in irregular war.

Primary sources most useful to the research behind this book include manifestos, manuals, and memoirs of irregular fighters, some of them leaders and some of them ordinary fighters. Other primary sources include official correspondence, reports, and records from the conventional military forces engaged in such conflicts, as well as those from the governments and international organizations involved. Secondary sources used in this book fall into three categories. First, academic works on specific conflicts and technical aspects of irregular war are used to frame the ideas presented in this study. The second category provides context, historical data, and broad detail on regional conflicts and crises. These first two categories include monographs, academic articles, and classic works on specific conflicts and regional issues pertinent to the concepts I seek to elucidate. Finally, a third set of secondary sources includes important works on political theory, political sociology, comparative politics, and relevant military studies. The data from these secondary sources is analyzed from different perspectives, with each yielding useful material for the central argument.

The plethora of data sets in conflict studies are not always comparable, whether the issue is time, type, outcome, or any number of other measurable items. However, the quantitative boundaries set by the number of insurgencies, civil wars, and revolutions that have occurred within bounded periods of time do illustrate a world beset by violence, much of which occurred in the context of irregular wars. For example, one study listed 181 insurgencies between 1946 and 2017, finding that insurgencies seeking overthrow were 43 percent successful, while anti-colonial insurgencies were 77 percent successful, and secession movements were only 10 percent successful.[47] The widely cited Correlates of War Project assessed wars between 1945 to 1997 and found that the 23 wars between states in that period produced 3.3 million battlefield deaths, while the 108 wars within states accounted for 11.4 million killed.[48] The numbers tell a story of wars inside a state occurring far more frequently and certainly more violently than wars between states, except the Second World War. An even broader study showed that although regular state militaries were responsible for the majority of the estimated 70 to 170 million casualties in the 250 wars counted between 1945 to 2008, states routinely claimed that it was the non-state actors who most often violated international humanitarian law by committing war crimes, slave-related activities, torture, and other gross human rights violations.[49]

The duration of insurgencies is also a challenge to define. One study found that most insurgencies lasted at least 10 years, with a minority tapering off at 16 years,[50] while another found that between 1945 and 1985 there were only 26 days of peace in the 150 wars or conflicts between states resulting in 20 million dead.[51] Irregular war occurs far more frequently than conventional war, with 120 of the 160 major conflicts fought between 1945 and 1990 clearly featuring this type of war.[52] For the U.S. alone, the military has deployed 253 times in operations that were not considered to be conventional from 1989 to 2001.[53] Despite all of this conflict, the United Nations executed only 13 peacekeeping operations between 1956 and 1988.[54] In this book, I will draw from the broad data that these and other quantitative studies deliver to underline aspects of my argument as I assess their findings and claims. The discrepancies resulting from the frameworks, methods, and interpretations in such studies are obvious, further illustrating that the field lacks a universally accepted point of departure for describing or explaining the phenomenon of irregular war.

The findings derived from the thick descriptions in select works have been most useful for the research behind this book. For example, a detailed study of the 1946 insurgency that produced the short-lived Mahabad Republic in Iran identified over 60 distinct tribes participating in the uprising, along with useful details about these varied participants, their methods, and their reasons for taking up arms against the state.[55] Similarly, a first-hand account of the Javanese insurgency that eventually ousted the Dutch from Indonesia in 1949 includes verbatim copies of telegrams sent between important insurgent leaders during the conflict, revealing in their own words why and how the insurgents fought.[56] A survey can also be useful for analyzing the onset, conduct, and conclusion of irregular wars, such as the case that produced a sovereign South Sudan in 2005.[57] These are helpful for their effective use of process tracing. Like these examples, each primary and secondary source used here assists in building the larger narrative necessary to test the stated hypotheses within my conceptual framework.

Structure of the Book

This book presents the novel concept of sovereign dysfunction within an innovative theoretical framework synthesizing the concept of justice as fairness, the functions of political institutions, and the process of group identity formation. The argument develops in a way that builds toward the claim that conventional forces employ the wrong paradigm for fighting irregular wars and that irregular wars are distinctly political before they are violent. More broadly, I argue that current models for participating in and explaining irregular wars are misguided at best and unsuitable at worst, leading to prolonged conflicts, unnecessary violence, and pervasive instability. To accomplish this, the book is separated into three parts of three chapters each.

In Part I, concepts described throughout the book are introduced, terms are defined, and conditions are set to analyze elements within a specific domain of inquiry. Chapter 1 establishes the frame of reference for the theory, including setting domains of time, subject matter, and the specific phenomena the theory applies to. I also briefly discuss methodology and describe the data assessed. In Chapter 2, I outline the theoretical framework used throughout the book. In Chapter 3, I present the argument for the novel concept of sovereign dysfunction as an independent variable affecting social orders, sovereign territories, and political-economic institutions in ways that often lead to irregular wars.

In Part II, I show that current theories and military concepts are inadequate for approaching irregular wars by addressing the flaws in their approaches, arguing their practitioners are fighting wars with illusory doctrines, and demonstrating how the evolution of conventional war theories are divorced from the causes and conduct of irregular wars. In Chapter 4 I construct a detailed argument against four problems in conventional war theories and military studies regarding irregular wars. These problems are rooted in scope, approach, bias, and character. The first problem revolves around a common confusion of warfare with war and theory with tactics. The second problem concerns what Carl von Clausewitz dubbed "methodismus,"[58] which is the tendency for military studies to favor doctrine and prescription over theory. The third problem follows what James C. Scott called "seeing like a state,"[59] or approaching the conflict as if the perspective of the incumbent state is the only acceptable lens through which to view a conflict. The fourth problem is one of character, in which scholars and practitioners become so bogged down in descriptions of specific conflicts that they abandon theoretical frameworks altogether in favor of unfalsifiable descriptions of conflicts in a vacuum. In Chapter 5, I demonstrate how current frameworks attempting to explain wars have generally developed out of a specific European milieu defined by tradition, convention, and rigid conformity. This chapter traces the historical development of war theories in France, Germany, Britain, and the U.S., while also showing how non–Western states such as China and Iran have adopted this narrow framework for their own war theories, strategies, and doctrine. In Chapter 6, I examine how states actually organize, train, man, and equip themselves for irregular wars. This includes a brief historical analysis of how current conventional methods originated and a demonstration of how those methods are designed for fighting organized state militaries and little else.

In Part III, I synthesize the original argument presented in Part I *vis-à-vis* the antithesis of conventional war theories I discuss in Part II. The result is a novel theory of irregular war, in which the functionality of the sovereign affects social, territorial, and political-economic components of the state, leading in certain cases to irregular wars. Chapter 7 explores the causes of irregular wars stemming from sovereign dysfunction and its interaction with social order, sovereign territory, and political-economic institutions using numerous cases from discrete milieux. This chapter shows how each case is path dependent and requires an examination of the causal factors over long periods that set conditions for irregular wars. Chapter 8 goes on to examine who fights, why they fight, and the ideologies behind which they unite. I employ French philosopher Émile Durkheim's concept of *anomie* and Arab sociologist Ibn Khaldun's concept of *'asabiyyah* (group-feeling) as two opposing positions in social belonging that have direct implications for the formation of insurgencies to rectify issues outside established mechanisms of the state.[60] Finally, in Chapter 9, I reflect on political scientist Kenneth Waltz's images of man, the state, and war, adjusting these images to the categories of insurgents, the state, and irregular wars. This chapter operationalizes the concepts presented throughout the book into a coherent theory of irregular war. I conclude by recapping the theoretical concepts presented throughout the book, suggesting areas for future research, and making a final admonition against using conventional theories of war to solve irregular war problems.

States often approach irregular wars with a military instrument so that their strategies tend toward force, suppression, and exclusion. These and similarly unsuited approaches beg for a theory commensurable with the phenomena it seeks to explain. Each case has a complex relationship between conditions and outcomes, in which causes and context matter more than technology and tactics. Revolutions in social orders, restructuring of sovereign territories, and transformation of political-economic institutions around the globe in the past few centuries provide a rich environment of cases from which to choose. Still, the current understanding of these and future conflicts remain insufficient to inform policy, prepare for war, and rationalize political violence. Militaries are prosecuting wars on false footing, to the detriment of all. Previous works on these conflicts have not addressed how structural changes to social orders, sovereign territories, and political institutions lead to irregular war when the condition of sovereign dysfunction is introduced, a gap I hope will come closer to closure after one considers the arguments outlined in this book.

Chapter 2

A Framework of Thought, Action, and Justice

> State building is the insurgent's central goal.—Stathis Kalyvas[1]

Useful frameworks for theorizing about irregular wars must accurately describe who is fighting, how they are fighting, and why they are fighting. Before doing so, a framework must explain the conditions leading to the onset of hostilities, the conditions that exist as hostilities progress, and the conditions leading to the conclusion of hostilities. This framework must then allow for an analysis of the effects of sovereign dysfunction across the variables of social order, sovereign territory, and political institutions. Once sovereign dysfunction has had sufficient influence over conditions, a leader may emerge to guide the people into the violence of resistance, rebellion, and revolution. But who will lead, and how can scholars best understand the processes at work once such a leader takes charge of an insurgency?

Two important frameworks can be used for explaining these phenomena, with one focusing on the ideology of the individual participant and the other focusing on the social structures involved in irregular wars. One may ask why a study of irregular war must use one theoretical framework dealing with the individual on one hand, and a separate framework for explaining the social orders, institutions, and states on the other? John Rawls provides an answer:

> The primary subject of the principles of social justice is the basic structure of society, the arrangement of major social institutions into one scheme of cooperation.... The principles of justice for institutions must not be confused with the principles which apply to individuals and their actions in particular circumstances. These two kinds of principles apply to different subjects and must be discussed separately.[2]

An excellent example of a framework for explaining the role of the individual in irregular war is found in Ali Shariati's 1979 lecture series compiled into the tract entitled *School of Thought and Action*. Meanwhile, John Rawls outlined a framework centered on justice as fairness in his 1971 book, *Theory of Justice,* which applies broadly to societies, institutions, and states in a useful way for explaining irregular war. It is worth discussing both as they provide the scaffolding for the theory that develops in the coming chapters.

A School of Thought and Action

Strands of conflict often evoke philosophical traditions bound up with them. For example, the class conflict following the Industrial Revolution led eventually to the Frankfurt School and its critical theory.[3] Similarly, revolutionary Shi'a Islam experienced an intellectually reflective tradition culminating with Shariati and his theory of man, ideology, and conflict, while Sayyid Qutb contributed to the development of militant Sunni Islam.[4] It could be argued that Shariati's *School of Thought and Action* was to the Iranian Revolution in 1979 what Rousseau's 1755 treatise *Discourses on Inequality* was for the French Revolution in 1789.

Ali Shariati provided a theoretical framework to explain the relationship between who leads, who fights, and why they participate together toward a possible violent death with no guarantee of success in the context of irregular war. Shariati's important works *Religion vs Religion* and *School of Thought and Action* both drew upon a post–Marxist critical theory while carefully following Max Weber's concepts of the state and Émile Durkheim's outline of society, all toward developing a framework for social revolution. Shariati followed Weber in framing the coercive power of the state, and Shariati agreed with Weber that secular and religious avenues for wielding a coercive power are legitimated through obedience to authority.[5] Durkheim recognized obedience to a leader as the source of moral authority in his idea of social consciousness, noting in 1912 that, "when we obey someone because of the moral authority we recognize in him, we follow his advice, not because he seems to be wise, but because a psychic energy immanent in the idea we have of this person makes us bend our will and incline to compliance."[6] The renowned social psychologist Stanley Milgram provided empirical evidence that Durkheim was right. Milgram found that "the key to the behaviors of subjects lies not in pent-up anger or aggression but in the nature of their relationship to authority. They have given themselves to the authority; they see themselves as instruments for the execution of his wishes; once so defined they are unable to break free."[7] Shariati provided a framework for examining

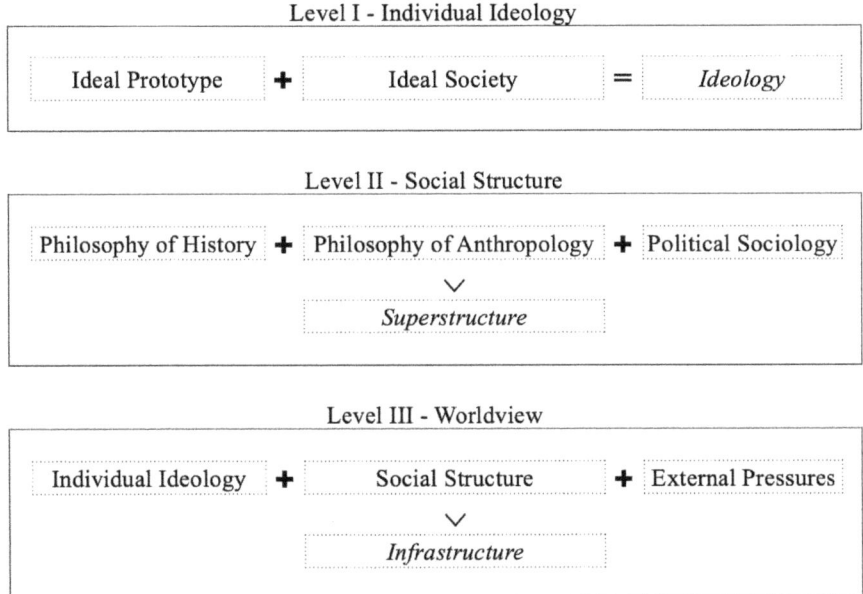

Figure 1. Individual Framework: School of Thought and Action.

the power such authority wields over the individual in society. Much of the raw material in Shariati's theoretical framework applies across conflicts, and it is therefore useful for a study of the phenomenon of irregular war. Figure 1 provides an adapted version of this framework suited for the current study.[8]

Figure 1 depicts the relationship between the individual, the social structure that shapes their behavior, and the worldview that individual holds. At the first level, *ideology* is formed through the combination of what the individual sees as the ideal prototype for a person and the ideal society in which that ideal prototype lives. At the second level, a *superstructure* is formed through the shared philosophies of history and anthropology that shape the social order, while the structures of political sociology accepted in that social order operate to establish rules of behavior. Finally, at the third level, the individual ideology from the first level and the superstructure from the second level combine under the effects of pressures external to the individual to form a *Weltanschauung* (worldview). This final integration of the individual into the structure of society to form a worldview forms the infrastructure of human existence as part of a social order.

Shariati's theoretical framework is useful for explaining the development of a *Weltanschauung* leading to insurgency and revolution. For

Shariati, the idea of *gharbzadehgi*[9] (Westoxification) acted as an adversarial counterpoise to the ideal society in the time and place in which he wrote.[10] This framework has three levels of analysis: the individual, the society, and the state or global order contributing to conflict, depending on the case. This parallels, but does not mirror, the three levels of analysis Kenneth Waltz used to examine the individual, the state, and the conflicts states wage with other states.[11] Waltz's images framework is particularly useful for general studies of war at the level of theory, but only wars between states. For irregular war, these images can be modified from the original levels of analysis of individuals, states, and the international system to individuals, groups, and wars within states. As Waltz noted, "The vogue of an image varies with time and place, but no single image is ever adequate."[12] The school of thought Shariati outlined is championed by "prophets of revolution."[13] Thought and action are separate categories of the same social force, splitting the authority that wields primacy over the individual into one group under the state and another under religion or, or in secular cases, an idea demanding devotion.[14] Like Shariati, Carl Jung differentiated between power vested in the state and in religious authority. Like Jung, Shariati was the intellectual heir to a deep history of philosophy of the individual and society.

Shariati's greatest intellectual antecedent was al-Farabi (870–950 CE), author of *The Virtuous Regime*. In this work, al-Farabi separated social orders into three groups: the wise, insightful leaders; their followers who act upon their faith in the sages; and the rest of society, the majority and least enlightened of all. Though al-Farabi carefully followed Platonic philosophy, he was the first to reconcile the religious ideology of Islam with the secular philosophy of Classical Greece. For al-Farabi, the best regime, the most virtuous, was one ruled by the diarchy of "the philosopher-king and the prophet-legislator,"[15] perfectly capturing the schools of thought and action Shariati described. Edward Said conceived of latent thought versus manifest thought, further breaking Shariati's thought into two important components: one beneath the surface, influencing behavior, speech, and *Weltanschauung*; the other deliberately expressed, transforming words to deeds.[16] Regardless of the divisions in Shariati's thought and action, ideology reigns supreme.

Shariati's theoretical framework is therefore principally ideological, with a particular focus on religion. Shariati identified two functions of religion.[17] The first is the priestly function, which is self-perpetuating and based in authority. The second is the prophetic function, which dictates morality and is based in social justice. Shariati gave a Hegelian character to these two aspects of religion: the first perpetuates hunger and the second satiates it.[18] For Shariati, their synthesis occurs in a revolutionary struggle for freedom from domination. In such a struggle, there are two outcomes in this process of social dominance, evident in his historical example of Sassanian Persia. In Shariati's view, the Sassanian aristocratic class subjugated the people through exploitation and coercion, while the priestly class used the more potent and enduring construct of religious legitimation to maintain their ideological authority.[19] This revolutionary interpretation of religious belief or secular ideology is not unique to Shariati, but his theoretical framework is a useful lens with which to view the organization, motivation, and agency in insurgencies and irregular war.

Other frameworks begin with similar observations but stop short of presenting a theory to explain the phenomenon. For example, one framework identified visionary and organizational leaders in revolutions, corresponding roughly to Shariati's thought leaders and actors, but did not delve into the processes endowing these individuals with their defining characteristics.[20] Another framework employed label pairs like "visionary and implementer" or "catalyst and champion" for the power dynamics of social groups.[21] Stepping away from academia for analysis can sometimes be helpful to explain how schools of thought and action affect organized social violence, as a voice speaking from the wars in Vietnam shows. According to Vo Nguyen Giap, the military leader of the Viet Minh during the Vietnam War, the primary purpose of political work in insurgency is twofold: "political education and ideological leadership."[22] Giap borrowed this from Mao Zedong who, calling explicitly for the separation of thought and action in China's civil war, directed that irregular forces in that conflict must divide their leadership into political and military wings.[23] Indonesian resistance movement leader Simatupang gave a useful metaphor for this division of leadership in a secular movement, describing the motor and rudder of "the ship of the state," in which separating one from the other "meant having a motor without a rudder or, conversely, a rudder without a motor."[24] During the Indonesian War of Independence (1945–49) against the Dutch, the *dwitunggal* of Sukarno and Hatta was a secular version of the duumvirate Shariati described.[25] Likewise, during the short-lived Kurdish Mahabad Republic in 1946, Mullah Mustafa Barzani and Sheikh Ahmad were the two components of Shariati's duumvirate, with the former the practical leader and the latter spiritual.[26] Although these and other authors, leaders, and participants were unfamiliar with Ali Shariati, his pattern of thought and action rings through all. This is why it is so useful.

John Rawls and the Theory of Justice

John Rawls provided a theoretical framework that can be used to measure the level of dysfunction within a state, whether in terms of the

Principles of Justice for Institutions	
Principle I:	All members of society have equal rights to equal liberties.
Principle II:	Benefits reach the disadvantaged while public offices are fairly and equitably available despite socioeconomic disadvantages.
Difference Principle:	If inequality may result, equal distribution should be sought.
Priority Rules of Just Social Order	
Priority Rule I:	Restrictions on individual liberty must improve the overall system of liberty in the social order. Those with less liberty than others must accept this disparity.
Priority Rule II:	The priority of opportunity growth must provide the greatest equality of opportunity to those with least access. Accumulation must be redistributed in a way that alleviates hardship.

Figure 2. Social Framework: A Theory of Justice.

social order, state sovereignty, or political institutions. This framework is "justice as fairness," explaining ideally functioning political institutions in well-ordered societies. Rawls argued his theory of justice was a social contract theory serving as an alternative approach to utilitarianism. Figure 2 outlines a theoretical framework of justice in ideal society used throughout this book in the context of social orders, political institutions, and the state.[27]

The social framework of justice as fairness in Figure 2 is divided into two operating categories: justice in institutions and justice in social orders. This framework explicitly links the categories of social orders and political-economic institutions that are dependent variables in this book. That linkage is affected by the dysfunctional sovereign. As the dysfunction increases, equality of rights and liberties erodes, usually in a zero-sum slide toward one group gaining rights and liberties at the expense of another. The ideally functioning sovereign equitably distributes these rights and liberties while also fairly distributing public goods so that upward mobility is possible for the most disadvantaged. Just social orders depend on an equilibrium between liberty and equality for all members of the society. As balance is lost, liberty, equality, or both become concentrated in fewer hands and hardship spreads to more members of society. Unjust institutions operating within unjust social orders deny members

Chapter 2. A Framework of Thought, Action, and Justice 27

of the society a means of addressing their grievances, prompting them to seek ways to adjust the system outside those structures. This external adjustment, if violent, manifests as irregular war.

In contrast to a dysfunctional sovereign, Rawls described the ideally functional society as one that is "a well-ordered society, one effectively regulated by a shared conception of justice, [while] there is also a public understanding as to what is just and unjust."[28] This conception of justice differentiates between theories of strict compliance and partial compliance.[29] Strict compliance describes the ideal society in which justice is perfectly aligned between the people and the institutions upon which they depend for fairness in everyday life. Partial compliance refers to the situations in which there is a dysfunction between the people and the institutions meant to distribute justice throughout society. Therefore, any study employing this framework must begin with the assumption that cases studied are partially compliant to some degree.

Although Rawls self-identified with Immanuel Kant, he did concede that ideals exist for theoretical purposes while actual societies exhibit the elements of partial compliance. This was hardly a novel distinction, as authors such as Thomas Hobbes and Georg Hegel had already done so.[30] It is because of this reconciliation of the real with the ideal that we can situate a theory of justice within the social contract tradition, directed by rational choice.[31] Rational choice frameworks can be used for examining collective action problems and for penetrating the "black box" decision-making processes of the state. Rational choice methodologies inadvertently promote avoidance of low-probability/high-impact events because they focus on the likely rather than the possible, an approach poorly suited to conflicts involving entities that are not bound by the obligations, stakes, and sunk costs of the state nor by the formal structure of games.[32]

As a subset of rational choice, game theory generally assumes individuals devise strategies for their behaviors following an assessment of the environment and other actors participating in the bounded system.[33] These games are categorized based on factors including the number of players, symmetry of the actors, availability of information, and cooperation between the players. Game theory models demonstrate that most interactive social phenomena occur in pairs due to the conservation principles of closed and semi-closed systems.[34] Irregular wars are no different, as they involve the process of bargaining between parties competing for limited stakes.[35] This type of bargaining activity is manifested through collective violence. Irregular war is one form of the coordinated destruction in which the conventional forces of an incumbent state receive significant challenges from organized irregular forces.[36] The lethal contest of coordinated destruction therefore includes irregular wars once begun.

Before the onset of violence, however, the conflict manifests as an attempt to leverage opportunity and activate mediation processes viewed as relatively productive of justice within an existing order.[37]

Using the "justice as fairness" framework, sovereign dysfunction becomes evident when injustice, inequality, and unfairness begin to overtake an unwilling population. Rawls observed that "one may think of a public conception of justice as constituting the fundamental charter of a well-ordered human association."[38] It follows that a departure from this charter may lead public opinion to view the system as unjust and, in some cases, ripe for violent transformation. The concept of justice as fairness allows for a sufficiently broad set of cases to be examined despite their external differences with one another because the framework places no special value on any one institutional structure or social order within a sovereign state if the principles and priorities Rawls described exist in some form within it.

The Functional Sovereign

Now that the structure of ideal justice in society and the individual's place in that society have been defined for our purposes, the ideal regulator of justice and of the individual must be examined. An ideally functioning sovereign state aligns its executive capacity, security guarantees, and public goods with the demands of its people. There is a common pattern visible in the formation of a state in which a sovereign eventually monopolizes enough force, real or imagined, to reach a point at which its functionality affects its own existence. Ingredients necessary for the state to reach this point include the following[39]:

1. A ruling entity monopolizes sovereignty.
2. That sovereign delegates power.
3. The delegation of power and transition into force by its agents is generally considered legitimate by those living under the aegis of such power.
4. The governing structure becomes an expression of social validation of the sovereign.
5. The sovereign gains legal jurisdiction over important elements of society.
6. The sovereign executes power and control through its agents over a physical space that no other sovereign may exercise simultaneously.
7. The social order under the sovereign is differentiated, enduring, and ideologically tethered to the state.
8. Membership is determined by birth, residence, or formal identity.

As the state drifts or is forced from an ideally functional to a dysfunctional form, opportunities emerge for factions to alter the fundamental social order, dissolve territorial control, and contest institutions comprising the bureaucracy of the sovereign all toward correcting the dysfunction within the system or creating a new system toward an imagined ideal, potentially leading to irregular war.

Given the frequency of this phenomenon, some tangible conditions of sovereign dysfunction tend to appear across cases. John Locke provided an important starting point for explaining and applying the conditions of sovereign dysfunction in his 1690 work, *Two Treatises on Government*. In this work, Locke described the process of revolution as legislative alteration, which, he argued, would lead to dissolution of a government but not of the power invested in society.[40] Locke arrived at this argument after explaining six forms of sovereign dysfunction[41]:

1. Arbitrary will is exercised in place of the law.
2. The sovereign abandons its charge.
3. The representatives or electors are changed without consent of the governed.
4. The government is prevented from assembling.
5. The government or legislature breaches its trust.
6. The people are given over to, or taken by, a foreign power.

These six conditions are not all-inclusive, but they do provide an important foundation for further developing the argument in this book concerning the causes and conduct of irregular wars. These conditions primarily deal with the sovereign, the electoral institution, and the people, so they only apply to governments in which power is shared between several individuals or factions in society. Hoarding power unevenly, such as in authoritarian governments, may instigate other conditions apart from those Locke described, but they are nonetheless apparent and generally consistent. Other conditions include interactions with the sovereign territory, political institutions, and the economy, and any of these could lead to political violence and irregular wars if sovereign dysfunction is sufficiently present.

Irregular war is a violent conflict between the people and the state in which the social order, sovereign territory, or political institutions of the state are threatened not directly by another state but by organized factions seeking to violently alter the structure of the state, replace it with another structure, or annihilate the existing state altogether. These organized factions may seek to circumvent the state to meet their needs, obtain security, or defend an ideal as the disconnect expands between the expected behavior of the state and the execution of its policies. They may seek such a change in three areas subject to sovereign dysfunction. The process at work can be explained in these violent eruptions by examining the independent variable of sovereign dysfunction as it affects the dependent variables of social order, sovereign territory, and political institutions.

The acts of altering, dissolving, and contesting structures of state power may occur bloodlessly or violently, whether votes are cast with ballots or with bullets. The premise of this theory of irregular wars is that these wars occur in the context of social orders, sovereign territories, or political institutions of the state depending upon the degree of dysfunction in the sovereign. Sovereign dysfunction is the disconnect between a state's *de jure* structure and its *de facto* behavior, especially in the liminal space between what the people demand, what the state promises, and what is delivered. Let us now define each of these three dependent variables of social orders, sovereign territories, and political institutions.

Social order is a structure of status, power, and authority that is generally accepted by its members having surrendered some autonomy in exchange for mediation and framing of values, norms, and beliefs. An intact social order provides predictable and just outcomes in social interactions. If disrupted, social constraints on violence, mistrust, and factionalism are removed as members operationalize their grievances through collective action, ranging from strikes and protests to secession and revolution for varying conditions and objectives. The theoretical element of the character of social orders points to the first hypothesis in this theory of irregular war: *States with weak political institutions and low centralized authority are prone to irregular war because grievances, power transitions, and social concessions cannot be processed through the current social order and instead reconciliation is attempted in a liminal space outside state control.*

Sovereign territory is the physical space over which a state defends its ultimate right to exercise legitimate force, administer justice, and extract resources from the people. An intact sovereign territory is characterized by uncontested control of the land it claims to administer, the capacity to govern outside the core, and a stable affiliation of periphery territory with the administrative nucleus. A disintegrated sovereign territory is characterized by lack of central government authority over areas outside core nodes, lack of security guarantees, and the inability of the state to defend key resources within the territory. Extending authority to these weakly governed spaces through repression generates ephemeral control. However, the long-term effects of those policies have paradoxical outcomes as they lack the inclusivity necessary to contain, address, and resolve grievances within the state, giving the aggrieved no other choice but to subvert, avoid, or destroy that system.[42] This theoretical element of sovereign territory constitutes the second hypothesis in this theory of irregular war: *States lacking centralized control over their territory are more susceptible to conflicts over contested geography, contributing to the onset of irregular war.*

Political and economic institutions are the official and unofficial frameworks for managing and regulating resources, activities, and changes within the political system to maintain its equilibrium. Rawls supplies us with a more formal definition of an institution:

> [An institution is a] public system of rules which defines offices and positions with their rights and duties, powers and immunities, and the like. These rules specify certain forms of action as permissible, others as forbidden; and they provide for certain penalties and defenses, and so on, when they occur.[43]

More simply, political and economic institutions are "the political constitution and the principal economic and social arrangements" of a society.[44] Elites have three recourses to address social unrest through political institutions: concession, repression, and democratization. The theoretical element of political institutions leads to the third hypothesis in this theory of irregular war: *Institutional processes that are sufficiently resisted, too weakly enforced, or outright rejected set conditions for irregular war as power becomes contested between the disenfranchised, the weak, and the oppressed against those perceived to control the exclusive yet sickly institutions.*

The Dysfunctional Sovereign

Now that the theoretical elements of the dependent variables have been identified, it is time to examine the independent variable: the sovereign. John Locke argued that the sovereign government cannot exist without a social order, distinguishing between dissolution of government and dissolution of society.[45] Jean-Jacque Rousseau defined this sovereign further by demonstrating how the people are the *sine qua non* of government when he declared that "the public person thus formed by the union of all other persons ... is now known as the *republic* or the *body politic*. In its passive role it is called the *state*, when it plays an active role it is the sovereign."[46] In Rousseau's view, sovereignty is "nothing other than the exercise of the general will" and the sovereign "is simply a collective being."[47] When the actions of the collective sovereign being become dysfunctional for the good of those who will its existence, a condition for irregular war exists. That condition, "the Kingdome of Darknesse"[48] as Thomas Hobbes put it in his *Leviathan*, is often rectified through actual or attempted social or political revolutions and their irregular wars.

Locke's justification for revolution is echoed in the American Declaration of Independence, in which Thomas Jefferson outlined the effects of a dysfunctional sovereign on a social order, whereby "...it becomes necessary for one people to dissolve the political bands which have connected them with another...." That declaration gave a litany of 27 grievances against King George III, closely mirroring the justifications given in Locke's *Two Treatises on Government*. Jefferson wrote that:

> ...whenever any Form of Government becomes destructive to these ends, it is the Right of the People to alter or abolish it, and institute a new Government.... Governments long established should not be changed for light and transient causes; and accordingly all experience hath shewn, that mankind are more disposed to suffer, while evils are sufferable, than to right themselves by abolishing the forms to which they are accustomed. But when a long train of abuses and usurpations, pursuing invariably the same Object evinces a design to reduce them under absolute Despotism, it is their right, it is their duty, to throw off such Government, and to provide new Guards for their future security.[49]

Ultimately, sovereignty is an expression of territory and people bound by authority. A well-established tradition in Western political philosophy points to a shared conception of what a good government means to its people. Rousseau described the functional sovereign as "bring[ing] together what right permits with what interest prescribes so that justice and utility are in no way divided."[50] The egalitarian social system in Thomas More's *Utopia* is one such ideally functioning sovereign and is not too far removed from the ideal system in Book VII of Plato's fourth century BCE work, *Politeia*. The former focused on ideal practicalities of society, while the latter sought a perfect justice. Thomas More articulated the benefits of the functioning sovereign thus: "But forasmuch as perfect concord remaineth and wholesome laws be executed at home, the envy of all foreign princes be not able to shake or move the empire, though they have many times long ago about to do it, being evermore driven back."[51] Locke's description of the functional sovereign included the three mandatory elements of law, justice, and power. Locke's ideal sovereign provides "the good of Mankind" and "the preservation of their property."[52] The functional sovereign acts as a "shackled leviathan," an equilibrium point between state dominance and social control of that dominance, precariously existing within the "narrow corridor" shaped by the unique system of norms exerting power over it.[53] By uniting utility of the individual and justice of the many, philosophers including Rousseau, Locke, and More anticipated Rawls in framing the ideal sovereign, legitimate government, and the stable social orders these create while their ideas demonstrate how the absence of these ideals may lead to irregular wars.

Violent Resistance and the Dysfunctional Sovereign

This theory of irregular war requires a framework built upon the human element and all its atoms of action and reaction, because, as sociologist James C. Scott asked, "How, finally, can we understand everyday forms of resistance without reference to the intentions, ideas, and language of those human beings who practice it?"[54] In its most basic form, the irregular force consists of political-coercive and informational structures that are often cast in some likeness to a state. This force is often described as an insurgency.

Competition, raids for resources, and organized violent conflict result when negative relationships exist between diversity, productivity, and economic interactions.[55] Irregular war is more likely when combinations of these elements lack balance. Peripheral areas have more irregular wars than central areas for the same reasons. When territory is highly aggregated, and the people are highly affiliated, less political instability and high internal security result. When territory is highly disaggregated and the people in the disparate parts are weakly affiliated, the state will likely face a violent challenger from within, if not also from without.[56] Figure 3 illustrates these theoretical assumptions.

The interactions of state productivity and social diversity affect how valued assets are stored, transmitted, and modified. On the ideal end, a

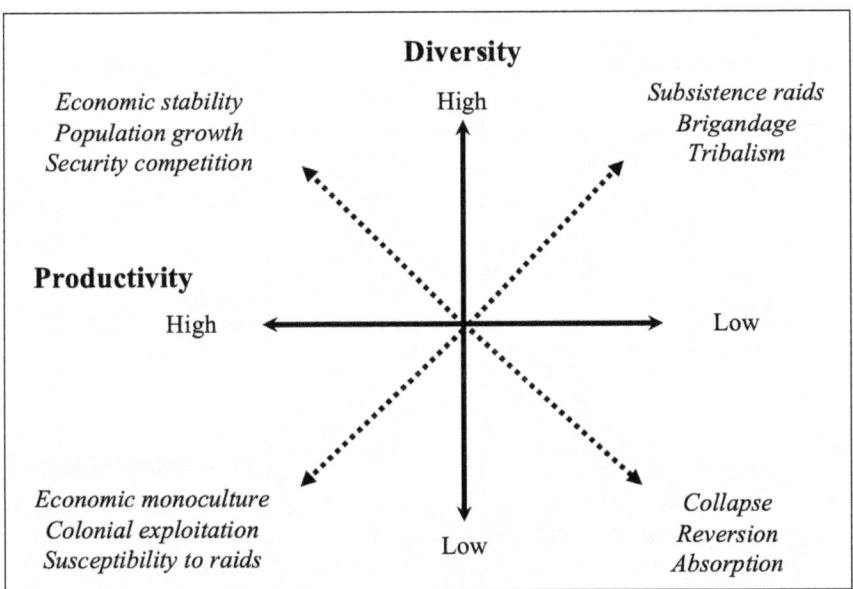

Figure 3. State Productivity and Social Diversity.

highly productive sovereign ruling over a socially diverse constituency is more likely than any other arrangement of these two factors to feature economic stability, maintain sustainable population growth, and engage in security competition rather than warfare with its peers. On the opposite side of the spectrum, a totally unproductive sovereign lording over a homogenous social group is most likely to experience collapse of the society, reversion to a more dissolute structure, or absorption by a competing sovereign with more capacity to dominate. In between these two extremes, a sovereign with a combination of high productivity and low diversity will likely experience colonial exploitation like in Congo, raids such as in Syria, and economic monoculture found in the oil economies in the Gulf states. On the opposite of this, sovereigns with high diversity but low productivity may experience subsistence raids as in Sudan, brigandage such as in Afghanistan, and the tribalism found in Jordan. While no ideal sovereign exists in practice, the higher the productivity and diversity of the state and its people, the less susceptible this bond will be to interruption by organized violent resistance.

The opportunity for insurgency grows as the sovereign moves further from the ideal. Carl von Clausewitz expressed concern over the peculiar nature of insurgency, something he termed *Volkskrieg* (people's war). Clausewitz noticed that in insurgencies, "the principal of resistance exists everywhere, but is nowhere tangible ... form[ing] threatening clouds from which now and again a formidable flash of lightning may burst forth."[57] Interestingly, Clausewitz reified the people in such a war, whom he described opaquely as a "nebulous vapory essence," treating them separately from the first leg of his well-known trinity of people, military, and government, an oversight later writers would repeat.[58] In fact, the people in irregular wars are neither one reified leg of a triangle nor are they a third of a "remarkable trinity." Instead, they consist of the irregular forces engaged in hostilities, those who somehow support or oppose those forces, and the general population, any of whom may flow across these roles or hold several simultaneously depending upon the circumstances.

An insurgency does not arise in a vacuum. It is simply a primary group forming with the specific objective of rising against an incumbent dysfunctional sovereign, often with the intent to change the social order, secede from the territory, or overthrow the state. Insurgencies must maintain certain structures and execute specific functions to either constitute a threat to an incumbent state or to have any hope of establishing their own. The group must construct a political identity, promote a popular cause, transcend rivals, and obtain sanctuary.[59] At the unit level of analysis, an insurgency forms when a group transitions from disparate social entities or groups into what is known in group dynamics theory as a primary group. That transition is often spurred by the corrosive effects of unresolved social forces unique to the context in which the group forms.[60] The primary group is uniquely defined through its interactions, goals, interdependence, structure, and unity.[61]

The attitudes of the current and potential participants are critical, and the insurgency benefits from crafting and exploiting a message to connect itself to people's needs.[62] That message is an essential element of the insurgency, both structurally and functionally. The model of rhetoric Aristotle constructed is useful for framing the way messages access attitudes. In this model, the pathways of communication are *ethos*, *pathos*, and *logos*. Each of these is a channel for reaching the listener, whether that channel is based on moral, emotional, or logical arguments. These each influence perceptions of authority in terms of reward and punishment, honor and shame, the internalization of the message through perceptions of credibility, and identification with the cause as a function of its perceived attractiveness.[63]

The exact structure and process of group formation continues to be debated, but some useful concepts are effective for the purposes of this book. One useful theory identifies five steps of group development: forming, storming, norming, performing, and adjourning.[64] Another way to look at the same process gives the steps of orientation, conflict, structure, work, and dissolution.[65] The rationally acting individual may surrender some or all of that rationality upon joining the primary group, with the individual adopting the primary group's perceived worldview as one's own.[66] Once the group is formed, the insurgency requires certain essential elements key to challenging an incumbent state, including local support, favorable terrain, funding, communication, supply, and external support.[67] The feeling shared among group members is a special category of these elements, as it permeates all aspects of the group. The group cannot survive without this shared group-feeling, or *'asabiyyah*,[68] while total isolation from the group produces anomie, its opposite.

'Asabiyyah and Anomie

Regardless of how the group is formed, the tightest bonds are built through a combination of primary group cohesion and identity assimilation. This process of cohering and assimilating is best captured in the Arabic term *'asabiyyah* (group-feeling/social cohesion). Ibn Khaldun used the term *'asabiyyah* in his 1377 magnum opus in conjunction with the term *ittiham*, which refers to embracing the *'asabiyyah* of the ruling elite.[69] In this sense, *'asabiyyah* is inherently both a social and political group-feeling. Some Western scholars have attempted to take credit for inventing this concept, such as one author's recent claim, terming it "groupness."[70] In contrast, those writing geographically and culturally near the birthplace of the term make no attempt to claim the concept, instead reinterpreting *'asabiyyah* simply as "ardent tribal solidarity" used to explain, for example, the success Muslims had in overthrowing the Sassanian Persian empire.[71] In general, primary group cohesion, or *'asabiyyah*, is formed through a combination of an ideological image projected from the leaders and a shared consciousness among the members who mobilize around that image, following the framework of Ali Shariati's schools of thought and action outlined above.

The idea of *'asabiyyah* applies across conflicts. Mao Zedong's *'asabiyyah* was the unity of spirit, which he considered to be "a powerful combat factor" that united ideology and solidarity of cause.[72] In the Latin American context, the post-conquest concept of *nepantla* (feeling-in-between) was a shared group-feeling among surviving Aztecs, not unlike what Mayans separately experienced during the Guatemalan War (1960–96).[73] The Guatemalan War produced significantly lopsided rates of human rights violations from the government against both insurgents and civilians. In that war, 89.7 percent of human rights violations were committed by government forces while only 4.8 percent of the human rights violations were definitively committed by the Leftist insurgents.[74] In all, 150,000 civilians were killed or missing. Certainly, this shared experience promoted a tangible *'asabiyyah* among those surging against the state during the conflict.

In a continuum representing the individual and group belonging, the totally cohesive *'asabiyyah* can be expressed visually at one end while an absolutely independent individual detached through anomie resides at the opposite side of the spectrum.[75] An ideal identity sits at equilibrium with society in the center between anomie and *'asabiyyah*. The relationship between these identity states is depicted in Figure 4.[76]

The spectrum of total isolation to total acceptance encompasses the entire experience of individual relationships with other people. The equilibrium of stable integration represents the integration of the ideal individual into an ideally integrated society. In that ideal case, all justice is equitably and fairly distributed, public goods are allocated where necessary, and the competing demands of individuality and belonging are in

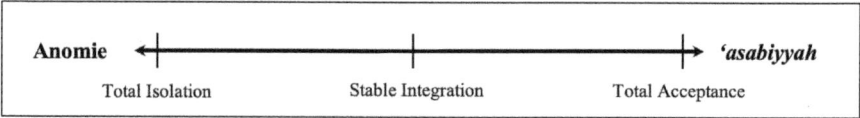

Figure 4. Individual-Group Belonging Continuum.

harmony. In contrast to this ideal, total anomie reflects complete isolation from a group, while *'asabiyyah* represents the total surrender of individual identity through the process of group assimilation. Although anomie and *'asabiyyah* represent two extremes on a single continuum, people find themselves more often moving toward *'asabiyyah* than toward anomie by virtue of our desire for social proof, shared identity, and belonging.

The concept of *'asabiyyah* is manifested in numerous close-knit groups, whether secular or sectarian, nationalist or anarchist, democratic or communist. For example, *tawhid* (oneness/monotheism) is an Islamic principle simultaneously describing universal submission and unity of the divine and serves as a fundamental tenet of belief for a quarter of the world's population.[77] From the secular perspective, the French *Union Sacrée* (sacred union), was a unification of otherwise disparate groups who banded together, or at least abstained from opposing the government, on the eve of the German invasion in August 1914. Institutionally, military forces such as the U.S. Marine Corps promote the feeling *esprit de corps* (group spirit) to deepen bonds between members when faced with shared adversity. These concepts of unity, oneness, and indivisibility frame the group rather than the individual as the ultimate quantum of force and identity, underscoring *'asabiyyah* as a necessary ingredient for a successful insurgent movement.

Chapter 3

The Structure of Irregular War

> Among diverse theories of conflict—corresponding to the diverse meanings of the word "conflict"—a main dividing line is between those that treat conflict as a pathological state and seek its causes and treatment, and those that take conflict for granted and study the behavior associated with it.—Thomas Schelling[1]

A useful theoretical framework for explaining irregular war can be synthesized from several disciplines in the social sciences. In this chapter, four parts of the theoretical framework required to approach irregular war are established. First, the general theory and philosophy of war are outlined to set the domain of conflict, whether nuclear, conventional, or irregular. Second, the argument is developed that social order, sovereign territory, and political-economic institutions depend upon the functioning of the sovereign. Next, the crucial distinction between irregular war and irregular warfare is made, being a necessary step to understand this theory and conflict in general. Finally, the structures of conflict often found in irregular war cases are addressed, namely insurgency and revolution. Once this structure is in place, the inadequacy of existing military theories of irregular wars will become obvious.

Theories and Philosophies of War, or "On Clausewitz"

Many scholars begin their studies of war with the ideas in Prussian military theorist Carl von Clausewitz's incomplete treatise *Vom Kriege* (On War), published posthumously by his wife in 1832. However, not all those later works agree with his original thesis, nor has every attempt approached a full understanding of the implications of his work. Clausewitz's detractors arose soon after his own writings were published. For example, Antoine-Henri Jomini, a rival officer and prominent theorist in

Chapter 3. The Structure of Irregular War 39

Napoleon's French *Grande Armée*, derided Clausewitz, at one point arguing that Clausewitz's "logic is frequently defective."[2] Clausewitz repudiated Jomini thus: "To see the whole secret of the art of war in a single formula, at a certain time, at a certain point to bring up *superior masses* was a restriction overruled by the force of realities."[3] Clausewitz despised the treatment of art and science in war given by the likes of Sun Tzu, Machiavelli, and Jomini, labeling their combined approaches inferior handicrafts. Indeed, Clausewitz felt a book on the *art* of war was useless, since in his view the study of war theory was a *science*.[4]

Clausewitz began his study with war's nature, theory, and strategy to frame the rest of his work. He limited his study to a certain form of war, a blend between *Kabinettskrieg* and Napoleonic warfare, noting "It is only for general, great, and decisive combats ... that this description [of the nature and character of war] stands good."[5] Clausewitz formed a paradigm by which the study of all war is still framed, yet many have not appreciated the limitations and scope of his work. The study of irregular war is no exception. Clausewitz placed the origin of irregular war in Europe in the nineteenth century, and his short chapter on the subject of "people's war" attempted to describe the potential military effects, uses, and causes of a force drawn up from the people.[6]

The concept of partisan wars, insurgencies, and other non-traditional conflicts have long been discussed. Notably, Jomini differentiated between partisan conflict and what he called "regular warfare," while Clausewitz considered light troops, partisans, and "irregular bands" as threats to conventional lines of communication, supply, and strategic flanks, such as at Napoleon's disastrous campaign surrounding the Battle of Borodino in Russia in 1812.[7] Early military-theoretical treatments of irregular war included the 1813 Prussian *Landsturm Edikt*, which Clausewitz helped write.[8] Clausewitz recognized the political and military components of such a mode of war that set it apart from the conventional war to which his magnum opus *Vom Krieg* is devoted. Those studying Clausewitz have often overlooked this distinction, instead fixating on the enticing expositions of violence and convention described earlier in the work.

Martin van Creveld serves as the archetypal example of a military theorist misinterpreting Clausewitz in the context of irregular war. Van Creveld fell into the trap of misreading a dialectic argument, grasping the opening statement of a premise, and being overcome by it before moving through the thesis, antithesis, and synthesis Clausewitz employed to deliver his points. For example, van Creveld was so consumed by the idea that war is an unconstrained bloodbath where there are neither laws nor human rights that he missed the rest of the concept toward which Clausewitz was methodically leading the reader: war is actually constrained by the state.⁹ Van Creveld also looked at war through a deterministic lens rather than understanding war as a social phenomenon. The tendency to view war and politics as an inevitable march toward historical development is narrow and Eurocentric, keeping tightly in step with widely-accepted conventional traditions prevalent in Western scholarship.¹⁰ Like those enamored with Clausewitz, many remain influenced by the idea of an inevitable, depressing, and fallacious march toward the end of history that Georg Hegel warned was on the horizon in his *Phenomenology of Spirit,* which he wrote as he witnessed France defeat his beloved Prussia on the battlefield at Jena-Auerstedt, not far from his chair at the University of Jena. It is interesting to note that both Clausewitz and Hegel produced their most famous works during the same battle that saw their conventional Prussian military forces utterly defeated by the superior strategic employment and *esprit* of the revolutionary French *Grande Armée.*

Because scholars continue viewing Clausewitz in a vacuum, it is unsurprising that van Creveld was left behind when Clausewitz went on to show that violence is tempered as it is necessarily subordinated to the reasoned influence of the state. Van Creveld's description of war as being "by the state, for the state, and against the state" was hopelessly narrow, serving at best as an idealized description of conventional war.¹¹ Van Creveld's work has had a Clausewitz-like effect on later scholars writing about wars, tainting their perceptions, distorting their analyses.¹² The baseline van Creveld and authors like him have set is no longer valid, if it ever was. A minority of voices have begun to dispute van Creveld's concept of so-called "nontrinitarian war," instead offering their own theories on the functions of war in the context of changing actors.¹³

Strategists have also faced difficulty in understanding Clausewitz. Famed military scholar B.H. Liddell Hart oversimplified Clausewitz, often quoting such lines as "blood is the price of victory," while ignoring the force Clausewitz used to promote the supremacy of defense over the weakness of offense.¹⁴ Like many who misunderstood Clausewitz, Liddell Hart missed the dialectic structure Clausewitz employed to arrive at each conclusion. Such self-professed students of Clausewitz instead grasped the first premise presented without realizing it to be part of a series of contradictory concepts to be synthesized later.¹⁵ Liddell Hart focused squarely on European history, taking a perspective that was deliberately Eurocentric, consciously excluding the Muslim conquests, Ottoman Empire, China, Persia, and even the Byzantine Empire.¹⁶ His history consisted of a line drawn from Pericles in Athens to Churchill in London, conveniently skipping anything that challenged upper-class British sensibilities toward history after the Second World War. His historical sources were narrower than Alfred Thayer Mahan's and only slightly wider than those Clausewitz referenced.

Practitioners of irregular war have had mixed results in their understanding of Clausewitz. Mao Zedong successfully drew upon Clausewitz and Vladimir Lenin to highlight the differences between the nature and character of war, while recognizing the limits of conventional armies in a people's war.[17] Clausewitz also influenced Soviet military thought: Soviet military doctrine, still used in Russian military circles, stated that "the essence of war is the continuation of politics by means of armed force,"[18] essentially paraphrasing the often mistranslated passage regarding the subordination of war to state policy in Clausewitz.

Others were not so astute, improperly framing Mao's use of the population in terms of Clausewitz's trinity of the government, people, and military.[19] The tenuous conflation of nineteenth-century Prussian military theory and a very different twentieth-century Chinese people's war indicates a conscious disregard for the incommensurability between the Napoleonic wars and the founding of the modern Chinese state in 1949.[20] Further, the Western concept of the cult of the offensive starkly opposes Chinese military theory, with the latter framed by a strategy of "active defense" in anticipation of wars of waiting for the enemy to make the first move followed by a decisive Chinese counterattack while the enemy was already deep in Chinese territory.[21]

Clausewitz's impact on U.S. military thinking is clear in military doctrinal publications. For example, a Marine Corps doctrinal publication entitled *Learning* explicitly parrots Clausewitz in stating that "the nature of war carries a combination of fear, uncertainty, ambiguity, chance, horror and, above all, friction that Marines must prepare to counter."[22] Three decades earlier, the Marine Corps released a Fleet Marine Force manual and a doctrinal publication both entitled *Warfighting*, listing the elements of war to include friction, uncertainty, fluidity, disorder, complexity, humanity, danger, and violence, adhering without deviation to the elements found in Clausewitz.[23] In this sense, we are dealing with warfare *per se* in which these elements could apply in any conflict regardless of cause, form, or parties involved.

Practitioners of traditional military activities have difficulty explaining or understanding irregular wars because their professional military education is designed for conventional, predictable, and mutually rule-bound adversaries. Starting from this position, analysis faces constant difficulty and competing ideas are hard to reconcile. This perspective has led to a belief that all war exists in an idealistic yet "'unstable equilibrium' of international politics and the rivalry of sovereign states for power, glory, and trade."[24] Some have challenged this view, or at least attempted to add nuance to it, differentiating between types of conventional war based on the enemy in a particular conflict, implying that the objectives of a given war may be revealed after examining its belligerents.[25] In this more nuanced view, there are three purposes of war: annihilation, benefit, and *gloire*. Another way to categorize the war aims of states is by analyzing motivations in terms of either territory or policy.[26] However, even the dominant theory of international politics, Realism in all its forms, is insufficiently structured to explain irregular war.[27] There are five prominent arguments scholars often use to attempt to explain these conflicts, but none sufficiently explain a rationalist argument for war. For Realists, the chief issue in interstate war is territory, though some argue instead that political institutions were most often the culprit.[28] A common perspective held since the early nineteenth century gives nine motivations for states to go to war; however, all these regard states fighting other states, or a state against itself, for reasons ranging from defending rights to "a mania for conquest," none of which actually apply for irregular wars.[29] If scholars cannot agree on the causes of purely conventional wars, then it is unsurprising that concepts in irregular war such as guerrilla warfare are frequently misunderstood.

The Cold War produced an interpretation of guerrilla warfare that followed the ideas in Clausewitz, in which this form of violence was "the extension of politics by armed conflict."[30] This was a clear attempt to modify the principle of subordination of war to state policy in the opening pages of *Vom Kriege*.[31] However, this understanding exists in the context of Marxist-Leninist conflicts, coloring the view of guerrilla warfare in a negative light, most notably by insisting that revolution follows a Hegelian trajectory to an end of history, or that insurrection would follow a necessarily historical path in respect to class conflict, such as that aroused through dialectical materialism. This attempt at forcing a square peg into a round hole elicited attempts to modify Clausewitz's definition of war in terms of the irregular wars raging globally during the Cold War.[32] Although many authors who framed their analysis from state perspectives in such conflicts as the Cold War and War on Terror misunderstood Clausewitz and irregular war, not all military theorists have.

Military theorists willing to break from the traditionalist view of war may draw three important conclusions from a reading of Clausewitz. First, war and strategy are distinct and therefore the basic theory of war Clausewitz described remains relevant in all contexts of war.[33] Second, the U.S. has a strategy deficit, i.e., it is good at fighting war but not as good at everything that follows once the bombs stop falling. Third, the so-called "American way of war" is unsuited for fighting irregular wars due to social, cultural, and traditional military aspects of such an approach to combat.[34]

Chapter 3. The Structure of Irregular War

In view of these three conclusions, among others, Clausewitz's framework was an important contribution to the strategic and theoretical understanding of war, though it was not, in the words of Clausewitz himself, "an algebraic formula for use on the battlefield."[35]

Ironically, the fields of political science and sociology have fared better in understanding Clausewitz than military theorists. One prominent view starts with the assumption that war should be a last resort rather than an easily exercised policy option.[36] In this view, the concept of distinguishing civilian from combatant never existed: as much as man is a political animal in the Aristotelian sense, he is also a nasty, violent, and brutish creature in the Hobbesian sense.[37] The sort of war fought between states is neither nuclear nor irregular; it is strictly conventional. Further, this conception of war must occur outside domestic politics, taking place instead as a contest chiefly among states rather than within them. That final point is where this common view diverges semantically with Clausewitz on war.

War is a cultural phenomenon, and each case is unique. There is a certain "cultural logic" to war, linking cultural products with the location, temporal and spatial, in which the war arises.[38] Following this cultural logic, the expression of violence in war is affected by time, space, location, and society. It is therefore unsurprising that a comparative study of six pre-industrial societies found that the individual culture of each was a greater factor in the character of their development than the process of industrialization itself.[39] Systemic analysis treats irregular war cases as components of a certain culture, functioning within what anthropologist Clifford Geertz called "webs of significance."[40] Humans inevitably produce these cultural webs through their interactions with one another, while established cultural webs dynamically interact with others in a never-ending dance.

This logic is usefully applied in a systemic analysis of international political economy and the conflicts arising from the discord in such systems. However, the level of analysis in studies of these conflicts focuses too greatly on state-level international politics rather than addressing wars within states.[41] Like the Realists, this approach emphasizes the uncertainty felt among heads of state along with the difficulties such uncertainties are bound to introduce when seeking resolutions.[42] The anarchy of the international system, in this view, forces military considerations into every economic and diplomatic decision, invoking all instruments of national power and revealing the tensions among them: they are simultaneously integrated and discordant. Such an emphasis on national power reveals the purely conventional conception of war that adherents to this view understand. But irregular war is no such affair.

Irregular War: A Social, Territorial, and Political Affair

The term *irregular war* was first attested in English as a form of war distinct from conventional war in the 1800 edition of Edwin Davies' popular tract *Other Men's Minds, Or, Seven Thousand Choice Extracts on History, Science, Philosophy, Religion, Etc.* A century and a half later, B.H. Liddell Hart distinguished between conventional and irregular war in its modern usage in the English language for the first time.[43] Yet Liddell Hart deviated little from the sense Davies gave to the term in 1800. The historical development of this term has been lost to scholars over time, as indicated by one prominent scholar in the field erroneously claiming that the CIA invented it.[44] In any case, Liddell Hart considered irregular war to be rooted in violence, a menace to authority, a form of war that breaks the rules. His wholly negative attitude toward irregular war led him no further in his discussion of it.

Liddell Hart's view exemplifies the attitude military theorists and practitioners often take toward irregular war. It is viewed as an aberration, a disease, an inconvenience that should soon end so that standard doctrine, drill, and ceremony can all resume once more in the comforts of tradition.[45] Scholars like Liddell Hart abhor a war that diverges from the long tradition of state against state, with common lamentations expressing frustration that "everything that can be called 'guerrilla warfare' has become a new military fashion or craze," despite its long history and persistence in all periods of human conflict.[46] Irregular war and indirect strategies have long been frowned upon, with scholars decrying the strategy Fabius Maximus used to defeat Hannibal during the Second Punic War (218–201 BCE), which fellow Romans considered "cowardly and unenterprising."[47] Winston Churchill, speaking while Secretary of State for War for Great Britain, shared his disdain for irregular war, calling it "miserable, wasteful, sporadic."[48] Traditionalists have characterized the period after 1945 as "an epidemic of armed revolutions" with outbreaks periodically silencing the good order of the Westphalian system.[49] To their chagrin "the disease continues to spread."[50]

As should be clear by now, most existing theories of war focus on conventional conflict and devote little attention to irregular war. A broad survey of major theories of war in several fields developed after the Second World War shows that studies of civil wars *per se* were an afterthought at best.[51] Attempts at studies of irregular wars often employ a triad model composed of conventional forces organized unconventionally, irregulars fighting state actors, and terrorism used to influence policy.[52] However, this forced analysis is hopelessly narrow and is, again, primarily focused on European history and state perspectives of war. More to the point, one scholar lamented that "the development of political complexity has attracted more scholarly attention than collapse, its antithesis" because the latter is a "difficult mystery."[53] As a result, separate streams of literature exist for the social, territorial, and political aspects of irregular war, and none provides a general theory adequately explaining this phenomenon. It will be helpful for the purposes of this study to examine those inadequacies so that we can identify what is missing and begin to see how a new theory could be framed.

Irregular War and Social Order

Irregular war is a social act, a group expression through a dialectical process within a state. In the literature of sociology, irregular war is often described in various terms, including insurgency, revolution, and civil war. Revolutions stemming out of the social order are unlikely when the sovereign is functioning as expected for the people it is shackled to serve.[54] In other words, no irregular war would occur if the coercive capacity of the state could retain its legitimated potency. If the sovereign is functioning, there is no irregular war. If the sovereign power is dysfunctional, conditions exist for irregular war. The essential elements of revolution are violence and necessity. In the presence of these elements, there is a relationship between ideology and terror in the expression of revolution.[55] Terror perverts revolution and is most strongly expressed by a singularity of potency against the many, rather than against the state.[56] Irregular war concerns the social interaction between the state and the people, who are sometimes called the masses, or, in older works, the crowd.

Gustave Le Bon's classic work *The Crowd* is a useful framing device for considering common attitudes toward the role of "the people" in irregular war, not for its staunch defense of the state, which it certainly is, but instead for its still-relevant flaws.[57] Le Bon inductively analyzed the characteristics of all crowds for his method using a Linnaean approach, which is unsurprising given the *zeitgeist* of late nineteenth-century France. He also focused on the condition of "becoming" rather than the crowd's trajectory toward its destination.[58]

Like Clausewitz, Le Bon's ideas were opposed immediately, notably by Russian aristocrat-turned-anarchist Peter Kropotkin. Kropotkin and Le Bon wrote at the same time and in the same place but viewed social groups much differently. In *The Conquest of Bread*, Kropotkin argued these groups could be taken advantage of as they engage in spontaneous activism. The group usually intends to adhere to the concepts of social justice and equality rather than "as a mob of savages ready to fall upon and devour each other."[59] Kropotkin also disagreed with Le Bon on the organism-like behavior of the thoughtless, collectively selfish mass and the intentionally selfish nature of competition within the human spirit of individualism, while also railing against Georg Hegel's master-slave concept,[60] presenting instead his own "mutual aid" theory of history. Despite rejecting Hegel, Kropotkin still relied on the dialectic method, revealed in his idealistic description of anarcho-communism, which he described as the synthesis of economic and political liberty. Conventional war theorists writing about irregular war tend to treat the mass, or "the population," in terms Le Bon promoted rather than those Peter Kropotkin or Ali Shariati used to describe the everyday people affected by war.[61] Le Bon's influence also spread among nationalist authoritarians of many flavors, such as Mussolini in Italy, Reza Shah in Iran, and Ataturk in Turkey.[62] In contrast, Kropotkin resonated with the disaffected urban labor force and the exploited rural peasantry in Europe.

Unfortunately for Le Bon, the people are not a hill to be taken; they are thinkers and actors. Whereas Le Bon saw social groups as unthinking bands of men reduced to their most animal form, Ali Shariati showed the masses to be a collective group of independent minds bound together in shared belief, a religion of revolution. In contrast to Le Bon's concept of the crowd as a mob, riot, panic, or trend, social movements are actually formed deliberately, with the view of transforming, amending, adding, or removing an issue affecting the group.[63] One must differentiate between "the people" and the degenerated form of "the mass," while also separating enduring institutions from relatively simple factions.[64] One of Le Bon's chief flaws in understanding the people was his failure to account for institutions. Political and economic institutions shape state behavior and affect societies in those states.[65] Several theories have arisen to explain this interaction, reflecting the numerous positions scholars haven taken regarding the role of the individual in social groups.

Some major theories have been proposed to explain revolutions, insurgencies, and civil wars, though none has yet fully achieved that aim. These include relative deprivation theory, modernization theory, and ideological explanations. For example, these theories may argue that irregular war is primarily caused by a combination of income inequality and environmental changes.[66] This variation on relative deprivation theory does have some merit in specific cases but is not broad enough to cover a wide range of cases nor can it probe the depths of the historical record. These theories lack sufficient explanatory power for addressing the "complex emergent process" of irregular war precisely because societies experiencing revolutionary situations are not passive monolithic entities, awaiting

Chapter 3. The Structure of Irregular War   47

some accidental spark of violence or being swept up unwittingly as some erringly imagine they might.[67] The optimism in modernization theory is also disagreeable for our purposes, as it is overly deterministic and clings to a Hegelian sense of history while ignoring contingent events, critical junctures, and institutional drift.[68] Arguments against modernization theory have helpfully affected the way coups, revolutions, and state failures are understood.[69] One useful conclusion emerging from these disparate theories is that whether states develop or decay depends on political, economic, and social systems.

States differ fundamentally from one another in part because of the institutions shaping their political economies. These differences may agitate or ameliorate conditions in the social order, potentially leading to irregular war. Many scholars agree there is a distinct difference between liberal democracies as defined classically and all other forms of government, though they diverge on classifications of those states that are not liberal democracies.[70] This difference is important because governments with intact institutions tend to avoid succumbing to irregular war.

Many have noted the tendency for liberal democracies to refrain from engaging in violent conflict with one another. In contrast, states that are not considered liberal democracies are more susceptible to internal conflict.[71] These latter states tend to serve as venues for insurgency, civil war, authoritarian repression over a majority, various forms of inequality, persistent foreign intervention, and a plethora of socioeconomic antagonisms that often lead to organized violence.[72] Importantly, various forms of inequality are linked to the socioeconomic conditions in those states.[73] This element of inequality forms a central focus of social approaches to irregular war studies.

Social approaches to irregular wars have long focused on the centrality of the peasant revolt within the context of agrarian societies experiencing successful, i.e., complete, social revolutions, such as in France in 1789, China in 1911, and Russia in 1917.[74] Causal-process observation is a helpful way to analyze these large-scale revolutionary movements, especially when the movements are overtaken by primary groups that had nothing to do with the crises they later come to use to further their agendas.[75] Lenin's overtaking of the general resistance to Tsarist rule in Russia is a prominent example of this, while the consolidation of power within Ayatollah Khomeini's circle in Iran after February 1979 is another.

Addressing social grievances in irregular war is a key process, a process Kropotkin described as "the study of the needs of man, and the means of satisfying them."[76] Several causal factors for irregular war are related to grievances, including the combination of lack of development relative to proximal example states, growth of population beyond the state's capacity to provide for such growth, and tangible unemployment.[77] Some military theorists reject this social grievance process, searching instead for in ill-defined and weak cultural explanation conforming to the proposition that it is not people and states that will be at war in the not-too-distant future, but cultures and civilizations.[78] Another way to consider the social problem without acknowledging the power of grievances is to separate social orders into two groups that manage legitimate violence: one works through a manipulation of political and economic institutions and the other works through competition. In this view, the first operates in the state in its "natural" form, limiting participation, while the second allows open access to institutions and participation, relative to the "natural" state.[79] Unfortunately, this conceptual framework is overly simplistic.[80] These social order approaches have yet to tie the varied explanations of violence together for intact political systems, nor have they broadly explained the violent breakdown of those systems.[81]

These gaps are important to theorizing about irregular war. For example, modernization theory formed part of the rationale supporting the 2003 U.S. invasion of Iraq, which led to a complex insurgency and state failure rather than fast-tracking to liberal democracy as originally intended. The American approach to counterinsurgency strategy was heavily influenced by a framework derived from modernization theory.[82] Employing modernization theory in counterinsurgency strategies implies a requirement for nation-building in the Western image of capitalism and democracy, without regard for which system the people in the target country actually prefer.[83] Similarly, relative deprivation theory was used in the context of American policies toward Latin American insurgencies and revolutions during the mid-twentieth century,[84] but this theory has since been discredited. The neorealist theory of international politics was useful but insufficient on its own to explain state behavior during revolutions, since most of the action occurs at the unit level of the structure rather than the "spare world" of neorealism focusing on the system level.[85] In particular, a key tenet of offensive Realism is that states prioritize survival above all else, including domestic political issues unrelated to survival but integral to the state's identity, such as the American values of democracy, human rights, rule of law, or impartial justice.[86] As these theories of the state and war are assessed, it becomes clear that none is sufficient for explaining irregular wars and the social orders from which they emerge.

Irregular War and Sovereign Territory

The process of state formation in the Westphalian system impinges on the territorial sovereignty of extant states. The goal in these conflicts

Chapter 3. The Structure of Irregular War

is to form a new state through secession, combination, or abstraction. In his study of Iran's post-revolutionary military development, Afshon Ostovar observed that "war and militaries [are] at the center of state formation."[87] States formed from irregular wars, such as South Sudan, Kosovo, or Timor-Leste were born through secession. States such as Yugoslavia, Indonesia, and the Soviet Union were born through combination. States such as Israel, the Islamic Republic of Iran, and the Turkish Republic of Northern Cyprus were created through abstraction, imposed upon a territory already considered wholly sovereign by other states, in these cases, Palestine, Pahlavi Iran, and Cyprus.

The concept of sovereignty as a quantum of force reserved in the state was anticipated in Jean-Jacque Rousseau's *Discourses on Inequality*.[88] Rousseau observed that "the outbreak that ends with the garroting or the dethronement of a sultan is just as lawful an act as those by which he, shortly before, held sway over the lives and property of his subjects. Force alone maintained him; force alone overthrows him."[89] Rousseau's statement reflects a specific kind of state, in which extractive institutions prevail under an absolutist government ruling by coercion, repression, and expropriation. Of course, not all states reflect this power dynamic. Nevertheless, Rousseau was describing the act of social revolution in a sovereign territory, the act of replacing the regime from the bottom up, rather than through foreign invasion, a coup of elites, or internal adjustments to political institutions. There is, however, a key difference between a coup and an irregular war. A coup aims to settle power within the context of the existing state; in irregular war the aim is to "smash the state" as Lenin put it, or to create "a party ... to set up a rival state structure."[90] In either case, Kurdish dissident Abdullah Öcalan considered the territorial state to be "the product of all kinds of internal and external warfare," offering instead the alternative of non-territorial sovereignty in his concept of democratic confederalism, which he described as "democracy without a state."[91] It should be noted that these and related views were considered so dangerous to the Turkish government that Öcalan was jailed for 10 years in solitary confinement as the sole prisoner on Turkey's İmralı Island prison.

The role of territory in shaping identity and political systems cannot be ignored. In his meticulous study *The Occupation of Enemy Territory*, Gerhardt von Glahn argued that "modern and practicable solutions must be sought for the problem of resistance groups, partisans, and guerrillas, despite the reluctance of professional military circles to grant recognition as combatants to enemy groups of this kind."[92] Irregular wars are not about states trying to win "hearts and minds"; instead, these conflicts are characterized by violence employed as a tool to affect social orders or assert dominance over disputed territory.[93] The highest levels of violence in civil wars exist in situations in which control of spatial areas is constantly in flux, while low levels of violence exist in areas of stable control, regardless of the actor exercising such control. However, those margins are not always violent.[94]

The isolating effects of terrain can maintain marginal areas as unique spaces with only tenuous connections to a state. This "special geographic dialect" maintained locally through geography and uniquely local social organizations acts to insulate those areas from the meddling hands of far-away elites.[95] A prime example of this in an urban geographical context was the resistance against French forces from within the Casbah in Algiers during the Algerian War of Independence. The Casbah was a densely populated, complex city within a city, and this physical reality facilitated resistance. Likewise, the social structures of the bazaar in Iran helped to topple the Shah in 1979. A theory of sovereign dysfunction benefits from the implications of how these liminal spaces exist in terrain, whether urban or rural, as the margins in territory may contribute to the onset of irregular war when the sovereign can no longer exercise power or force over those places.[96]

Other considerations of sovereign territory are often overlooked, including climate change and human terrain. Conventional military planners ignore the effects of climate change on populations susceptible to irregular wars, though climate is certainly a primary component of sovereign territory.[97] Human terrain is another important factor that can exist in a liminal space outside central control. In the case of Yugoslavia, there were important differences between the Chetniks who failed and the Partisans who succeeded.[98] The chief difference was not superior military training, special knowledge, or superior ideology. Instead, the difference lay in leadership, specifically political acumen rather than naiveté, and mastery of human terrain.

Irregular War and Political-Economic Institutions

Like war, political domination has a nature and character.[99] The Realist school contends that war is all about "the struggle for power,"[100] but this ignores the fact that most humans in history have been governed, have not significantly contested that governance, and do not themselves govern. Humans seek some form of authority and the resulting security assurances, predictability, and resource guarantees such a bargain provides. When this arrangement is threatened, war is likely. Submission and consent have the same result: obedience is rendered to the state.[101] Political domination alone does not lead to irregular war, but a combination of

Chapter 3. The Structure of Irregular War

grave political and economic issues within the state can lead to civil war and state failure.

The contest between elites and the masses is largely centered around economic issues, with such issues sometimes leading to transitions into or out of democracy.[102] In such conditions, economic stagnation equals civil war and starvation as there is no incentive to innovate, invest in the market, or maintain capital, while extractive institutions lead to state failure.[103] Factions may resort to violence if they cannot participate in the political system and its constituent political institutions are not inclusive.[104] At this intersection of economics and institutional capacity failures, dysfunction results amid a breakdown of statecraft, appropriation, and control, leading often to irregular war.[105] The frequency and similarity with which such conflicts have arisen between large segments of the working poor against expropriating elites has contributed to the appeal of Marxism-Leninism to many movements in the last hundred years.

The Marxist-Leninist interpretation of political-economic structures in the state is important to consider for understanding irregular war despite its declining relevance after the fall of the Soviet Union. Because economics involves what Thomas Schelling described eloquently as "divisible objects and compensable activities,"[106] Marxists and other economically focused sociologists or historians artificially narrowed their observations of human interactions into categories of labor and property, often exclusively. Marx recognized problems, or antagonisms, between classes he observed in the web of conflicts in continental Europe during the period following the French Revolution of 1789 through the events of 1848. For Marx, "The whole history of socialism and the political struggle was bound to disappear,"[107] a view later adopted by Vladimir Lenin following the Bolshevik Revolution in 1917.

Lenin saw the transition from state to non-state as a synthesis beginning with the dictatorship of the proletariat followed by its withering away into a classless society. Lenin noted that this classless society did not require total destruction of the state but only enough departure from the state system so as to wrest sufficient energy from old power structures for the next phase. James C. Scott dubbed Lenin a "high-modernist" revolutionary, having the top-down planning of convention, utility, and parsimony poorly disposed for a grand revolution demanded by the masses.[108] For Lenin, the vanguard party was to the mass what a general staff was to an army.[109] Lenin viewed dialectical materialism as the solution to political, economic, and social problems before him, much as Le Corbusier saw high-modernism as the solution to the unplanned, all-too-human Paris of the nineteenth century.

Lenin remained within the framework of history that Hegel

contextualized while also emphasizing the centrality of class, labor, and property Marx promoted. However, Lenin saw the synthesis of these systems of antagonism and class differences resulting in the total abolition of class. Lenin believed that the sovereign state sprang from irreconcilable antagonisms within classes.[110] Lenin also identified three phases in the Marxist class-based history of political power: a state existing to empower an exploitative minority against the working class consisting of proletariat and peasant; a violent revolution wresting power from the minority elite, establishing a "dictatorship of the proletariat,"[111] in which all trappings of state are destroyed and then centrally reconstituted to suppress exploitation; and finally, abolition of class and state, a "withering away." This view colored many movements across the globe in the twentieth century, even if their core beliefs were not totally aligned with Leninist thought. States rushed to counter those movements even if the puzzles states sought to solve through warfare were not well understood.[112]

The Conduct of Irregular War

War is the concept of organized violent conflict among competing factions, and this is distinct from the details of how wars are conducted. In the context of violent conflict, the employment of strategy, tactics, and operational art using instruments of power to achieve an objective is called warfare: it is the conduct, method, or way of the war. Conventional U.S. military doctrine distinguishes between war and warfare in a similar manner: "Warfare is the mechanism, method, or modality of armed conflict against an enemy. It is 'the how' of waging war. Warfare continues to change and be transformed by society, diplomacy, politics, and technology."[113]

From the perspective of military theory, war and warfare are distinguished partly through the philosophy of conflict ethics.[114] However, this philosophy centers firmly on the state and the attendant subjective moral arguments for agents of the state such as soldiers, police, and surrogate forces. These and other doctrinal and philosophical treatments of war generally do not see an irregular form of war, rather placing this irregularity at the level of war's conduct, i.e., warfare. If irregular war is a form of organized violence, then irregular warfare is the conduct of the irregular war. Irregular warfare concepts must therefore describe who is a belligerent and how they are fighting.

Irregular warfare has gone by many names, with guerrilla warfare among the most common. It is worth briefly surveying the concept of guerrilla warfare in its usage among military theorists and practitioners

Chapter 3. The Structure of Irregular War 53

so we may refine what is meant when using the term. Denise Bindschedler defined guerrilla warfare as "only a form of combat, which very generally is characterized by the fact that it is conducted by dispersed and mobile groups who avail themselves of surprise attacks, ambushes, sabotage, and generally avoid battle in the open field."[115] Another perspective holds that guerrilla warfare pits multiple players into a competition of endurance and will, where discrete military victories hold little value.[116] Vietnamese military strategist Vo Nguyen Giap defined guerrilla warfare as "The form of fighting by the masses of a weak and badly equipped country against an aggressive army with better equipment and technicians," a definition that basically defines the Vietnam War.[117] Another influential definition of guerrilla warfare described it as "an irregular war carried on by independent bands."[118] This last definition is important because it relates to the German concept of *Bandenbekämpfung*, which translates to "combating bandits" in English. The idea that guerrillas are simply bandits or, in recent wars, "unlawful enemy combatants," permeates Western military thought almost without challenge. The successful urban sniper John West observed that "experts in the field—and even the soldiers themselves—generally don't understand guerrilla warfare because they are trapped in a world dominated by conventional militaries, and their conventionally-minded leaders."[119] Few definitions of guerrilla warfare place a violent contest over the legitimate use of force at the center, where it belongs. Despite the inextricable link between politics and guerrilla warfare, most theorists and practitioners focus on tactics and targeting.

It is perhaps interesting to note here that the modern targeting method of "find, fix, and finish" is hardly novel. This targeting model was hailed as a revolutionary innovation for counterterrorism operations in the first decade of the twenty-first century, but it is identical to Samuel Griffith's concept of "location, isolation, eradication," which he employed to describe counter-guerrilla operations in his 1937 commentary on Mao Zedong's *Yu Chi Chan*.[120] While some have insisted that counterinsurgency operations should not use insurgent methods, Samuel Griffith countered instead that "the tactics of guerrillas must be used against the guerrillas themselves."[121] In any case, Griffith appeared to favor something like the British approach to the Malayan Emergency, while most others seemed to favor early American efforts in Vietnam that have proven to be largely ineffective in strategic terms.[122] The important issue in this debate is not what tactics should be used but instead that one should recognize counterinsurgency as an operation, not a strategy.

In irregular war, the act of employing the force is irregular warfare. The first inklings of a distinction between irregular war and irregular warfare arose in the 1970s amid important semantic discussions of the distinct concepts of small wars, *guerrilleros*, *Kleinkrieg*, partisans, *yu chi chan*, and so on that are often erroneously lumped together.[123] Rather than adhering to a "preferred style in warfare,"[124] irregular war theory must inform strategy, and the subsequent conclusions of strategy and operational art resulting from this careful understanding must frame the engagements in which states, or any other organizational structures, commit forces to irregular warfare. But how is this warfare structured?

Structures of Conflict: Insurgency and Revolution

Motivations to join an insurgency are not principally about greed but grievance. Costs for non-participation can be as high or higher than joining the insurgency. Social scientists have applied the wrong paradigm for understanding why individuals decide to join an insurgency despite the high costs, high risks, and low probability of individual gain.[125] Collective action and free-rider paradigms are inadequate for studying civil war and insurgency, because those paradigms do not adequately factor in risk, violence, and contingency.[126] The key question to ask in a study of structures of conflict is: Why are the people driven to fight?

Let us first examine an influential yet flawed perspective exemplified by French *para* and Algerian war veteran David Galula. Galula's 1964 book on counterinsurgency is an influential reference often used to approach this issue. It is required reading at numerous military academies, professional military education programs, and reading lists. After 2006, Galula's 99-page book became required reading for those attending the U.S. Army Command and Staff College.[127] Galula argued that revolutionary war was protracted, ideology was powerful, and revolutions were ultimately political. The debate among military theorists regarding the causes, character, and conduct of insurgencies can be framed briefly in a discussion of Galula's work, which one doting scholar described as "remarkable," "the best exposited" work on counterinsurgency in the last hundred years, reaching "the highest peaks in a whole mountain range of studies on irregular warfare."[128] Galula identified four causes of insurgency: social, economic, racial, and artificial.[129] This incomplete list reflects Galula's own experience and *Weltanschauung* as a French soldier fighting insurgents in colonial and post-colonial conflicts.

Following Galula, military theorists often favor four motivators for an insurgency: integrity of borders, political systems, sources of authority, and policies affecting society.[130] These are attributed to the rise of insurgencies during the 1960s, such as Maoist interpretations of

Chapter 3. The Structure of Irregular War

Marxism-Leninism. This was certainly one form, though not the dominant one, even in the examples of groups in Indonesia and Palestine that do in fact feature some of these motivations. This view lumps the war for independence in Algeria together with the Cuban experience, two completely different systems of conflict. Insurgency and revolution are not so easily distilled into what Galula identified as orthodox and shortcut patterns, applicable as they may be for some Marxist-Leninist cases.[131]

Galula adopted the "saturation point" theory for identifying the culminating point in an insurgency originally developed by Cypriot insurgent leader Georgios Grivas.[132] However, Galula inverted the theory for his own state-centered approach, turning Grivas on his head.[133] Galula saw diminishing returns for a growing insurgency due to relative forces applied by the counterinsurgent force to combat them. However, this saturation point theory emerged from the unique insurgency in the Cyprus Emergency (1955–59), and the man who developed it, Grivas, was ultimately defeated.[134]

Galula defined insurgency and counterinsurgency in terms of revolutionary war, attempting to remove these activities from a particular political framework, such as communist versus reactionary or republican versus royalist.[135] In contrast to Galula, the recent development of the "selectorate" theory in the field of political science presents a more convincing argument that the essential objective of an insurgency is to transform the insurgents and the group they represent from an excluded faction to an included one.[136] However, the reality is more complex still. This complexity is revealed in structural approaches to the conflict. There are nine structural elements that manifest uniquely in irregular war: organization, technology, logistics, direction, doctrine, decisive battle, soldiers versus warriors, allies and accomplices, and segregated versus integrated conflicts.[137] Argentinian revolutionary Che Guevara used a simplified structure, identifying principles, organization, and conduct as the central pillars of his own *foco* theory of irregular warfare.[138]

It is helpful to delve into Guevara's theory a bit to understand how a particular resistance movement generates novel concepts while coopting elements of existing philosophies to suit their unique needs. The *foco* (focus) theory centers on the guerrilla movement as instigator of a revolutionary situation and is a synthesis of Guevara's experience in the Cuban Revolution and his own interpretation of Mao's *Yu Chi Chan*. Because *foco* theory seeks to create revolutionary conditions rather than exploit extant grievances, it is flawed in its idealism. The theory favors movement from rural to urban settings, and not the reverse, while favoring geography over popular support. In a *foco* uprising, guerrillas must create a revolutionary situation *de novo* rather than participate in one already underway.

Guevara and others applied *foco* theory in Bolivia, Argentina, South Africa, Congo, and Angola, all with generally unfavorable results. The Brazilian resistance leader Carlos Marighella wrote his own theories on the subject in his 1969 *Minimanual*, which stood in contrast to Guevara's theory. Rather than advocate for the periphery-to-core flow of activity in *foco* theory, Marighella instead promoted an urban setting for the conflict.[139] Both Guevara and Marighella, though espousing opposing strategies, belonged to the milieu of post-colonial Latin America, a place very different from the *bled* in which Galula found himself in Algiers. From a sociological perspective, there are three issues shaping the *Weltanschauung* of the participants in irregular wars in post-colonial Latin America.[140] These are the sovereign states, third parties intervening to occupy, and the identities of racial or ethnic groups participating in such conflicts. Within these violent clashes, a subset of labor conflict is a recurring theme defined by land tenure disputes, elite exploitation of the labor force, and repression. These examples of theories developed organically within the Latin American contexts of Argentina, Brazil, Cuba, and elsewhere in the region emphasize the philosophies and goals of the insurgents rather than the states they resist during irregular wars and even in successful revolutions. There is substantial disagreement over what actually constitutes a revolution, however.

The early years of the U.S. provide an illustrative case of the long debate over what constitutes revolution and how one may categorize it. One common approach is to view only so-called "grand" revolutions as authentic, such as China's in 1911, Russia's in 1917, and Iran's in 1979.[141] This is still a nebulous distinction that leaves more questions than answers. Which conflict in American history, if any, was of the "grand" type? Was it the Revolutionary War (1775–83), the ratification of the Constitution in 1789, or the American Civil War (1861–65)? To answer this question, one must distinguish between two main forms of revolution: political and social. The first seeks transformation within a society subject to a political system while the second seeks to reallocate political power within that system. The difference is related to the location of the transformation, whether it is society that transforms to suit politics or politics that transform to suit society. For example, an argument could be made that the American Revolution was not a collapse of government but instead a transfer of authority.[142]

Following this distinction, one may argue that the American Civil War and French Revolution in 1789 were social revolutions, while the American Revolution and the ratification of the Constitution were of the political type. Following this logic, England's Glorious Revolution in 1688 and Japan's Meiji Restoration in 1868 were political revolutions.

Chapter 3. The Structure of Irregular War 57

Furthermore, revolutionary war is but one act in a revolution seeking transformation. Benjamin Bush observed this in 1787, as he saw the difference between the American Revolution writ large and its violent component in the American Revolutionary War, with the latter having become a common but incomplete synecdoche for the greater drama with all its acts, many of which were political and diplomatic rather than solely violent.[143]

Despite their merits, these arguments tend to treat the state as a black box existing in a vacuum, reducing their applicability. Importantly, Neorealist and rational choice theories of revolution have long been absent from this debate, despite scholars now recognizing that revolutions cannot occur without states, with the Peace of Westphalia in 1648 acting as the *terminus post quem* for such conflicts.[144] The very structure of the sovereign state forms the necessary substrate within which revolutionary movements arise, regardless of the declared motivations, composition, or rates of success of these movements.

The case of the rapid rise of Da'esh[145] in Iraq in 2014 demonstrates that shifting from one sovereign to another can be dramatically violent and is not always marked by a gradual drift through a punctuated equilibrium. Instead, irregular wars like the one in which Da'esh thrived are most often marked by near-unconstrained violence that quickly and tangibly transform political systems, if only temporarily.[146] The Da'esh example can be viewed as the successful exploitation of violence and time to achieve transformation of a political system rather than viewing the conflict in terms of state failure. However shocking and rapid the rise of Da'esh was, not every irregular war follows this recipe. In contrast to Da'esh, cases generally exhibit two ways the state can fail the people before revolution is likely given a protracted decline in sovereign territorial control, gradual evaporation of viable economies, and the gradual collapse of political institutions. First, the state becomes the perceived culprit causing the people's grievances.[147] Second, the state's institutions must effectively exclude the people to a degree that the risk of violent opposition is worth the costs to seek change from below or outside the state, rather than through existing institutions. Revolutionary wars and states are therefore intimately connected.

It is helpful to examine the connection between revolutionary wars and states from an international relations perspective to evaluate how states react to such conflicts within a neighbor's borders. Revolutions affect security competition among states in the international political system by destabilizing balances of power, causing shifts in alignments, calling agreements and legal frameworks into question, and creating space for states to attempt to maneuver into more favorable positions.[148] This international relations perspective is incomplete, however, as it centers on foreign policy considerations with revolutionary governments without considering how or why these governments were born. When bringing in those latter considerations, a framework begins to emerge indicating that revolutions are both social and political in their representation of popular rejection of established political ideas.[149] Samuel Huntington defined revolution as "a rapid, fundamental, and violent domestic change in the dominant values and myths of a society, in its political institutions, social structure, leadership, and government activity and policies,"[150] hinting at the difficulty and danger involved in such processes. This difficulty has long been known, as Machiavelli wrote in 1531 that "one must consider nothing more difficult to handle, nor more doubtful of success, nor more dangerous to manage than to make oneself the head of the introduction of new orders."[151] The "new orders" Machiavelli referred to can be categorized into two different baskets of conflict, with revolutions in one basket and other forms of domestic unrest in the other. Huntington usefully grouped "insurrections, rebellions, revolts, coups, and wars of independence"[152] in one basket, while revolution is something else and belongs in its own basket. This is because the first basket of conflicts only modifies elements of the overall system while leaving other major elements intact. It is important here to distinguish between changes of government occurring outside the normal political process on one hand and complete revolutionary transformations of political and social orders on the other. Those transformative events are "grand revolutions," such as the French in 1789, the Russian in 1917, the Chinese in 1911, the American in 1776, the Iranian in 1979, and so on.

This difference is observable when contrasting Marxist-Leninist revolutions, such as in Latin America, against largescale militarized popular uprisings such as the Ukraine Insurgent Army during the Second World War.[153] Indeed, some revolutions in twentieth-century Latin America were the final critique of labor exploitation in the banana industry, representing a unique cultural logic.[154] Causal mechanisms in these contexts are related to consolidated authority and exploitative economies centered around extracting labor and land possession.[155] Writing in the Communist context, Vladimir Lenin observed how the character of war changes when it becomes a violent social uprising, noting that "there are kinds of wars, including revolutionary wars."[156] These wars to which Lenin referred are social revolutions, and they are quite rare because of the high cost, greater participation requirement, and lower certainty of success.[157] The existence of this clear distinction has done little to prevent a tendency among observers to conflate social and political revolutions, leading to a synecdoche trap much as military theorists have done with war and warfare.

PART II
———

The Synecdoche Trap

CHAPTER 4

Problems of Scope, Method, Bias, and Character

War requires theory while warfare requires strategy, tactics, and operational art. Irregular war and its accompanying conduct, irregular warfare, may appear exotic and counterintuitive to students of war having been fully steeped in the ideas of Clausewitz, Jomini, and Mahan, or later authors.[1] Those later authors would be excluded in an academic work on irregular war if not for the outsized influence their writings exert on the most powerful Western militaries, not least of which is the U.S. military. Influential military writings in the last half century have often been more sensational than scholarly, tactical rather than strategic. The prominent military commentator David Kilcullen took a subjective approach to the study of war, for example refusing to use the word "mujahidin" to describe Afghan insurgents because he did not want to "cede to the enemy the sacred status they crave,"[2] an alarming embrace of bias inappropriate for an academic study of counterinsurgency. Kilcullen is considered by many to be "today probably the most influential terrorism and insurgency analyst in Washington. His work ... has informed the evolution of much U.S. counterterrorism strategy in recent years."[3] Kilcullen co-authored the 2006 U.S. Army and Marine Corps doctrine on counterinsurgency. Max Boot, another influential author unabashedly proud of his bias, was an advisor to General David Petraeus in 2010 in Afghanistan and an advisor to the U.S. military in Iraq after the initial 2003 invasion. Kilcullen wrote the irregular warfare section of the 2006 Quadrennial Defense Review, while Boot testified before Congress on war policy issues. These and other authors tend to attempt to bolster their credibility by mentioning that they have physically set foot in a minority of the countries about which they write. However, merely being present in the theater of major operations is an insufficient criterion for expertise, though these and many others do not shy away from attempting to frame it as such in the introductions to their works. If this criterion was sufficient, every destitute villager present in hamlets contested between the Viet Minh and the U.S. military would have emerged an expert on the insurgency in Vietnam. The American losses in Vietnam in 1975, Iraq in 2011, and Afghanistan in 2021 clearly indicate that simply "being there" is not enough to endow a person with the vision necessary to see, understand, explain, and succeed. Instead, these confused observers see an alien cultural landscape and remain bogged down in the problems of their calcified perspectives.

Many have used irregular *warfare* as a synecdoche for irregular *war*, creating four problems. The first problem is one of scope, in which scholars conflate tactics with theory. The second problem is one of method, in which prescription is offered over explanation. The third problem is one of bias, in which authors "see like a state." The fourth problem is one of character, in which authors drown in description rather than focus on facts. These four problems represent the key challenges faced in any attempt at formulating a theory of irregular war, and they must be examined to avoid repeating such errors.

Problem One: Scope

The formulation of irregular war theory and irregular warfare strategy is hampered by the scope in studies by military theorists and practitioners. For these, scholarship produced during interwar periods tends to overrepresent the previous conflict, reflecting on what went wrong and idealizing ways the next war will somehow be either a radical departure from the last or an offshoot of it.[4] Many military theorists have clung to the minutiae of terminology, chronology, and one-sided analyses. Little is done to address the substantive differences in terms like "low-intensity conflict" and "low intensity war."[5] The resulting lacuna in military theory where irregular war is situated is rarely confronted. Scholars often lament that the U.S. wars in Iraq (2003–11) and Afghanistan (2001–21) lack sufficient "honest reflection," yet they fall short of conducting the kind of analysis necessary for understanding similar conflicts such as the one into which France launched Operation Barkhane in Northwest Africa.[6]

Unfortunately, that honest reflection is often lost in the intervening periods between conflicts. In 1960, a French officer fighting in Indochina noted that "modern weapons can help conquer land; but [guerrillas] conquer people."[7] Mao Zedong brushed aside the conventional soldier's reliance on the capturing of land as a measure of military success, instead stressing the importance of a battle over time, human terrain, and the individual character of all people subject to the conflict. Mao's focus was not on quick victories in battle or the employment of the best weapons.

Instead, his focus was on the entire situation and environment in which the conflict was fought while actively resisting the "mechanical approach to the problem of war" that preoccupied the West and its allies.[8]

In any case, many Western military theorists have placed themselves inside of a box rather than thinking outside of one. Clausewitz recognized the flaws in conventional military thought during his own time as he witnessed the crystallizing effects of a sacred rigidity in European military doctrine. He railed against what he called "the force ... [and] the trammels of fashion," viewing with disdain the many treatises circulating among the military elite of his day which were full of pithy aphorisms hanging like "small crystalline talismans" but were really just "trumpery rubbish."[9] This trend of failing to objectively link ideology with political violence continues unabated, most recently in the context of violent extremist organizations such as al-Qaeda, Da'esh, and the Turkestan Islamic Movement.[10] There is frequent reference to the "Darwinian process of combat" in military literature, as though capability and action alone thin the ranks and produce better leaders, which may be true only at the tactical level and only then in just a few cases.[11] However, the higher up the chain of command one looks, the more strength dependency becomes the determining factor of success rather than strategy, tactics, or operational artistry within the surrogate force leadership. Strength dependency is the form of dependency state sponsors engender in their surrogate forces during either irregular wars or unconventional warfare in conventional wars. Historical analysis often falls short in assessing non-state groups, for example with one study failing to acknowledge the Chechen role in the successful rise of Da'esh up to 2016, while others have erroneously claimed the precursor to this group arose in Iraq, Syria, or Jordan despite its birth in Afghanistan.[12]

Experts remain transfixed by a hypothetical spectrum of lethality and survivability of forces actually engaged in conflict rather than the elements of conflict itself.[13] This transfixion represents a confusion between the elements of tactical action. That is to say, these experts attempt to assign their findings somewhere in either the purpose, method, or end state of some conflict as though the non-state actors were following the same mission-type orders upon which Western conventional forces rely at the tactical level. However, much of their findings usually ring true only at the methodology/employment level of *warfare*, rather than the purpose/intentionality level of *war*. Further, claims such as "There is nothing intrinsic of state status during war"[14] completely ignores the potency that such pro-state frameworks as Geneva, Montevideo, Hague, and Vienna possess, nor that of NATO, the United Nations, or treaties states wield over other states in their conduct, framing, and expectations in war, even irregular ones. Indeed, irregular wars exist because the state system exists.

Many military theorists writing after the attacks of 11 September 2001 tend to focus solely on counterinsurgency, using this operation as a synecdoche for irregular war as a whole, claiming the cases they studied were from "a selection of warrior-scholars from different nationalities," when in fact they are usually European, American, or Israeli military officers.[15] Attempts to summarize current theories of irregular war and the practice of irregular warfare only serve to highlight how the mainstream of military theorists instead fall victim to their own observation that "scholarly understanding about this type of warfare is also problematic."[16]

Problem Two: "Methodismus"

A useful war theory eschews doctrinal prescription to avoid the "methodismus" against which Clausewitz warned.[17] "Methodismus" refers to the tendency to make doctrinal prescriptions rather than attempting to solve theoretical problems. Practitioners have begun to question whether current military "science" is actually scientific, with a minority concluding that the American military has instead tended toward scientism, a form of "methodismus."[18] The Army Design Theory is an example of this unscientific approach to solving military problems, in which irregular war is completely avoided from a theoretical perspective.[19] Indeed, even during the Vietnam War, irregular war was "completely overlooked by the military planning staffs of the Western world" yet the war in 1950s Vietnam was "only 15 percent military and 85 percent social and economic" despite the Western response to employ a military "that thought in terms of towns and roads [against] an enemy that thought in terms of people and countryside."[20] This inappropriate military response to a social problem is unsurprising, given one view that the state is essentially "a militarily structured entity."[21] A critical defect affecting current irregular warfare doctrine is that Western militaries are manned, trained, and equipped to provide forces for traditional military activities along with doctrine and training to support that primary mission at the expense of non-traditional or "special" activities.[22] This defect stems out of the broader and widely embraced understanding of war from which military strategy, operational art, and tactics derive. It was not until 2016, for example, that Congress finally created a permanent statutory mechanism for supporting, *inter alia*, irregular forces against insurgents. This was a full 15 years after the "War on Terror" began and nearly four decades since the group that would become Hezbollah killed 241 Marines in a suicide bombing in Beirut.[23]

American irregular warfare doctrine and its academic proponents use a method to understand insurgency that has its intellectual antecedents in Prussian cameral science.²⁴ This approach favors centralized authority, interpreting data in terms of the state, and ultimately leads to reifying "the people" and alienating insurgents. Such studies are flawed by oversimplification and a narrow vision. For example, whereas one recent study in the cameralism tradition identified only three types of irregular forces, Mao Zedong saw at least seven.²⁵ Cameralists relegate irregular war to a "substitute for conventional war," arguing that those using associated techniques are "the angry, the frustrated, and the weak."²⁶ Another frequently-consulted volume framed Mao's highly effective ten principles of guerrilla warfare as "simple rules designed for beginners in warfare."²⁷ However, in irregular war, those adhering to Napoleon's dictum that "victory goes to the largest battalion"²⁸ are trying to use the theory of gravity to describe quantum mechanics: it is the wrong set of equations for the phenomena at hand. Instead, theory must treat a broad set of cases and explain their phenomena rather than focusing on the last war, which is a single case selected on its dependent variable. The common error of "methodismus" in the literature of irregular war distills most works into studies of operations and tactics while reifying the insurgency. Irregular wars are lost on these studies.

Focusing too much on doctrine and prescription leads to shortfalls and weaknesses in the findings and implications of studies suffering from "methodismus." For example, one such study measured counterinsurgency success in conventional military terms for the French war against al-Qaeda in Northwest Africa, finding that "the French had successfully prevented the jihadists from taking over the country and killed or chased a great many of them out."²⁹ However, the forces remained at a stalemate in the short term. Ultimately, France withdrew, and al-Qaeda's now–battle hardened offshoots expanded in the long term. True, it is difficult to define when an irregular war has begun or, perhaps more challenging, when it ends.³⁰ To the detriment of sound analysis and the prevention of future wars it is easy to declare them over even if they are not.

Some place the military into a vacuum or study only one side of the conflict, and this usually means the conventional side. The results are thin and biased despite the sweeping implications these studies should provide given the way they are described as deep histories of insurgency, raiding, and banditry. In one such study, out of 18 case studies, only two concerned non–European leaders, and all cases occurred in a period of only two centuries.³¹ In another, the "paradigm armies" concept was imagined as a heuristic for measuring military effectiveness at any point in history without actually accounting for historical differences between time, place, and case.³² A study of Arab military effectiveness, including Hezbollah and Da'esh, attempted to defend the concept of a "dominant mode of warfare," in which capacity, technology, and resources could be used in an ideal way to reach desired military end states, such as victory, stability, or regime change.[33] These sweeping, artificial frameworks are inadequate tools for understanding why specific insurgencies develop, how they perform, and what the outcomes of the conflicts may be.

The comparative method is sometimes used to understand the factors at work in irregular wars, but these often frame conflicts in terms of conventional military doctrine and operations. In one attempt, an organizational theory was proposed for understanding the differences between British and American counterinsurgency operations in Southeast Asia.[34] Basic comparison was used in a similar attempt to explore the impact of culture on combat effectiveness in Arab militaries and non-state armed groups.[35] No such analysis is conducted in comparative detail for adversaries who are not conventional, with the possible exception of nuclear war theories.[36] However, the latter have been designed to dissuade the use of these weapons rather than to address irregulars *per se*.

In reality, irregular war has been a fact of life across the globe in every cultural and social context since the idea of the modern state was born. This is true whether one places the origin of states in the Italian republics, the Ottoman regional sultanates, or the Peace of Westphalia. The concept of a bounded political and social entity exercising ultimate authority over a territory could have begun in the European milieu of merchant city-states or statelets like Venice, Florence, and Palermo, or in the Ottoman context with Aceh, the Maghreb, and Yemen's interior. Indeed, they did in all these cases. Florentine statesman Niccolò Machiavelli used *il stato* (the state) in its modern sense for the first time in an enduring and meaningful way in 1531.[37] Still, some authors ignore this broader historical view, opting instead to apply their own "methodismus" to the problem.

The neoconservative pundit Max Boot is an archetype for "methodismus," and his influential book *Invisible Armies* is a powerful example of all the flaws this problem entails. Boot's tactical attraction to a strategic problem was buoyed to his belief that irregular war is less about political theory, economics, and the long tendrils of history, and is instead about "security, police work, and intelligence gathering."[38] His vision of success is defined by the Malayan Emergency and the Huks in the Philippines instead of thoughtful, objective consideration of the real mechanics at work in a broad range of cases.

Boot's work on countering insurgencies from the perspective of a Western occupying power from 1810 to 2010 was misleadingly clothed as a volume about irregular wars. Following a traditionalist, Eurocentric approach to history, Boot recast European history into a synecdoche of the total history of war, cramming 17 centuries of global human conflict into about 50 out of the 700 total pages in his monograph. Boot devoted the rest of his study to the post–Napoleonic period of European history, with few exceptions.[39] Boot's Eurocentric perspective resulted in identifying Britain, France, and Prussia as the "gunpowder empires" and failing to discuss the actual empires in this category, of which there were only three: the Safavids, Mughals, and Ottomans.[40] Further, Boot inaccurately identified the Caucasus as "home to obstreperous tribesman, mostly Muslims"[41] in the nineteenth century when in fact most were Orthodox Christians except at the coastal fringes of the Caspian Sea. In all, Boot presented about 50 cases, nine of which definitively occurred prior to the Napoleonic wars. Of the other 40 since 1800, only one did not involve a Western military power as a belligerent. The categorical method he employed was based on his own subjective definitions of "victory" and "defeat," two terms ill-suited for irregular war. The folly of using conventional categories such as these in a war for existence rather than one of treaties or tactics is immediately obvious. Studies such as this exemplify major problems with existing literature on irregular wars and irregular warfare, including the bias of seeing like a state.[42]

Problem Three: Bias

Irregular war scholarship is often penned by conventional military practitioners and is written from the point of view of the state. This kind of writing is "state-sponsored and often amounts to a narcissistic exercise in self-portraiture,"[43] as James C. Scott put it. Some have consciously attempted to work outside this bias while remaining bound to it. For example, one scholar discussed high- and low-capacity regimes in terms of collective violence, but this was a one-sided conception of power exertion and extent that is not as useful for irregular war contexts when the non-state forces are given equal analytic attention to that rendered unto the state.[44] Such bias introduces an important weakness into risk assessments for future conflicts, in which "a perceptual incentive structure"[45] is reinforced through the bias, stymieing accurate predictive capacities. Missed opportunities, increased risk, weakened predictive power, and excluded probabilities that would be otherwise important to risk calculation all follow from this weakness. The bias is exemplified in the commonly held perspective in mid-century French approaches to irregular wars that the objective in these wars was first and foremost "to control the populace"[46] rather than address their grievances or adjust policy. In fact, this bias seeks to preserve the status quo at all costs, eschewing morality and refusing to acknowledge that grievances are real and contribute directly to the growth of an insurgency.[47]

The French perspective influenced U.S. military thought soon after the U.S. began major combat operations in Vietnam in the late 1960s. French interrogators fresh from their failed counterinsurgency operation in Vietnam arrived at Fort Bragg, North Carolina, to teach their methods to the U.S. Army ahead of their deployments to Vietnam.[48] The U.S. was not alone in being influenced by flawed French methods in countering insurgencies. The Hornbeam and Damavand Lines Iran used to suppress the Dhofar Rebellion (1963–74) in Oman were not unlike the Morice Line France employed in 1957 in the Algerian War, both in form and mismatch to the types of conflict in which these were used.[49]

In contrast to the French methods employed in Algeria and Indochina, German Luftwaffe interrogators took a softer but ultimately more effective approach. After the Second World War ended, Horst Barth, chief of the fighter interrogation section in the Luftwaffe described the ideal interrogator as "A man who fights without weapons, fences without a sword, fights only with his brain. He must be a very curious man."[50] Hanns Scharff, another Luftwaffe interrogator, likened the job of the interrogator to one "building gigantic jigsaw puzzles by fitting together carefully and cleverly the plain and pell-melled pieces of information. Their work may be likened to the sagacious construction of the missing parts of a complicated mosaic."[51] While U.S. interrogation doctrine eventually shifted away from French "hard sell" interrogation methods and toward the much more successful German Luftwaffe "soft sell" interrogation techniques by 2008, French influence remains strong in U.S. doctrine for counterinsurgency, security forces assistance, foreign internal defense, and counterterrorism.

American military doctrine for irregular warfare draws heavily from perspectives that have been discredited yet continue to be promoted.[52] Because of the biases inherent to derivatives of the failed French colonial model, the Pentagon's definition of irregular war is problematic and exclusionary, while ideas framed as doctrinal concepts are really just operational frameworks.[53] These doctrinal publications represent neither theory nor strategy, though the problems they identify and the possible solutions available are not new. Charles Thayer first recommended establishing an irregular warfare command in the U.S. in his 1963 book *Guerrilla*, but two more decades fraught with hard lessons passed before this became a reality when U.S. Special Operations Command was activated in 1987.

Authors of U.S. military doctrine are self-aware of this problem, though they do little in attempting to solve it. This awareness is apparent, for example, in a statement in the U.S. Army Field Manual on counterinsurgency, which states that "doctrine focuses on how, while national policy focuses on what."[54] Crucially, neither doctrine nor policy are drafted to

Chapter 4. Problems of Scope, Method, Bias, and Character   69

ask "why." Meanwhile, the Chairman of the Joint Chiefs of Staff deferred the weight of theory further down the line by proclaiming that "policy is not strategy."[55] Few in the strategy realm wish to touch policy while policymakers shy away from strategy formulation, leading to an unintentional incongruence between policy and war.

Academic studies of irregular war are heavily biased in favor of states, conventional wars, and organized militaries. Instead, one may be better served employing a "neutral observation-language"[56] to observe, measure, and analyze irregular war, rather than employing charged language like "terrorism," "fanatic," and "criminal" to describe the subjects of a study. There are five forms of bias in studies of civil wars, including bias of partisanship, politics, conceptualization, selection, and over exaggeration.[57] The most important subjective problem in irregular war studies is "assuming those who combat guerrillas are inherently moral and that insurgents are inherently amoral."[58] Irregular war scholars would do well to heed Émile Durkheim's observation that studying a subjectively repulsive phenomenon does not equate to condoning it.[59]

One prominent example of this bias through semantics is the use of the term "North Vietnamese Army" for the Vietnamese fighters opposed to the U.S. intervention during the Vietnam War. This title was an American invention sanctioned by General William Westmoreland, who did not want to validate North Vietnamese sovereignty in the title of the North Vietnamese military. In reality, this force was called the People's Army of Vietnam. Westmoreland directed that the People's Army of Vietnam would be "a term that could not be used in print" because he felt that the popular resistance to U.S. forces were "not 'an army of the people.'"[60] This practice returned in 2014 with the intense debate on how U.S. policymakers should refer to Da'esh. These are distracting semantic debates that only expose the subjectivity of those forming and executing policies, focusing on the symptom (a violent insurgency) and not the disease (the causes of the conflict leading to the insurgency).

A narrow, incomplete corpus informs American irregular warfare doctrine.[61] This literature is dominated by authors who have all promoted state-centered and often anecdotal colonial and post-colonial Western perspectives.[62] These perspectives continue holding a monopoly on irregular warfare studies, despite their tight focus on the conduct of counterinsurgency operations rather than assessing insurgency causes, which is a "somewhat puzzling" practice.[63] Irregular war studies framed in terms of preserving the incumbent state and destroying the insurgency are commonplace, often emanating from the most influential voices in those disciplines.[64] In perpetuating this approach, they fail to address important questions in the conflicts assessed. This irresistible preference for preserving the state at all costs is a major contributing factor to state involvement in destructive, messy, and unnecessary wars that those states fail to understand.

There is a strong tendency in conventional military thought to force irregular war concepts into convenient, albeit incommensurable, conventional molds. For example, in an assessment of Iran's influence over Shi'a militia groups in the Middle East, one analyst described these groups as "very well organized under [Iran's] command," who "see themselves not as a loose assortment, but as a single army with a very clear structure and hierarchy," and that there were "many mini Hezbollahs" in Syria.[65] Inaccurate, poorly articulated claims such as these ignore history and impose the convenience of convention when none exists. Instead, the greatest commonalities these groups share are their enemy and their religious beliefs.

Some notice the "remarkable aversion to unconventional tactics"[66] in the American way of war combined with a preference for overwhelming violence of action, affinity for technology, and an air of moral exceptionalism borrowed from America's early political history.[67] It is therefore unsurprising that American authors deride the supposedly unsophisticated tactics of, say, Shi'a militia groups, while in the same breath they begrudgingly concede that "Shi'a militias have often punched above their weight in combat with Arab and even foreign militaries."[68] Indeed, the most advanced special operations forces have removed none of these groups from anywhere in the region, and their operations have only served to increase the latter's legitimacy and boost recruiting efforts.

It is important to distinguish here between conventional forces in irregular wars on one hand and non-state combatants on the other. The special operations forces of state militaries are really "irregular regulars,"[69] especially when employing unconventional warfare and operations behind enemy lines or other denied areas. NATO countries are not alone in using these special operations forces for these activities. The Soviet Union doctrinally understood unconventional warfare to include special reconnaissance, sabotage deep behind enemy lines up to capital cities, *maskirovka* (military deception), inciting discord in the enemy's rear areas, espionage, assassination, cutting enemy communications, and pathway defeat of nuclear weapons.[70] Iraqi military doctrine in 1991 similarly tasked special operations forces with deep reconnaissance, unconventional warfare, unusual tactical engagements, and vertical envelopment.[71] North Korean unconventional warfare doctrine among its Special Purpose Units includes direct action raids in enemy rear areas, deep reconnaissance, sabotage, and support to insurgency.[72] These tasks are all included, with minor variations in name only, in the current U.S. Special Operations Forces core activities. Despite states wielding these non-traditional capabilities, military journalist Robert Taber wisely observed in 1965 that "a

Chapter 4. Problems of Scope, Method, Bias, and Character 71

spotted jungle suit does not make a United States Marine a guerrilla [and] ... commando troops are not guerrillas."[73] Instead, for Taber, the guerrilla was "an armed civilian ... a political partisan ... whose principal weapon is ... his relationship to the community, the nation, in and for which he fights."[74] This accurate depiction is a far cry from General Westmoreland's refusal to view the Viet Minh as who they were: armed civilians waging a political war against a state.

French soldier Roger Trinquier gained notoriety for his brutal methods of torture developed in Vietnam and then exported to Algeria after France's failure at Dien Bien Phu in 1954. Despite witnessing and perpetuating these failures, Trinquier was unmoved by the evidence that the French approach to irregular wars at the time was totally ineffective. Reflecting later on those wars, Trinquier continued to maintain that irregular war, "like classical wars of the past, will definitely end only with the crushing of one of the two armies on the battlefield, or by capitulation of one side to the war aims of the opponent."[75] This echoes the conventional perception that matured after Dien Bien Phu, where Trinquier personally experienced the overwhelming defeat inflicted upon the French army by Viet Minh irregulars and the subsequent Geneva Conference ceding North Vietnam to the Viet Minh the same year—hardly a reprise of classical wars of the past.

Like Trinquier, French general Hubert Lyautey led a scorched earth colonial campaign across a North African population, this time in Morocco. Beginning in 1903, Lyautey developed a plan to systematically eradicate Moroccans resisting his campaign, drawing up maps that he called "smallpox charts" so that he could more easily cure the land of the disease of its inhabitants. Again, following the admiration of other French colonial officers, some American military theorists consider Lyauty's *tache d'huil* (oil spot) strategy in French Morocco to be "the greatest [counterinsurgency] doctrine of all time."[76] Still more attempts to study counterinsurgencies continue omitting the human factor from their analyses, for example focusing on the organizational characteristics of conventional military institutions in counterinsurgencies, but not on the enemies or host populations those militaries face.[77] These studies are misleading because they purport to address insurgency while instead discussing ways conventions help or hinder state military forces without regard for the causes of the conflict or analyzing competing hypotheses for the potential outcomes.

It should now be apparent that unbiased literature on irregular war is lacking. Although many non-state groups are fighting for complex reasons, their motivations are often couched in the simplest terms, if at all. One study bluntly concluded that all Islamic extremists everywhere fight out of a desire to "revert back to the questionable glories of the twelfth century,"[78] ignoring the nuances of Salafism, Wahhabism, Shi'a/Sunni issues, the effects of globalization, land theft, forced displacement, and many others. Analytically, this bias contributes to a failure to address the obvious path dependent variables at play such as foreign occupation, illegal settlements, nuclear weapons, extractive economies, political repression, and so on. Biased terminology is employed in these works, with examples such as kidnap versus capture, guerrilla assault versus raid, and strike versus attack. One or the other is used depending on whether the actor is a state or a non-state group conducting the same act.[79] In other words, a military can "successfully raid a terrorist hideout" while insurgents "shamelessly attack a Coalition base." The same act, yet the adjectives and noun choices change the perception of the action and start the reader's inference of events from a position of bias. For example, an academic study of the French war on al-Qaeda in Mali in 2014 described a violent engagement one group of fighters carried out against a group of Malians. Was it a French "raid" or an insurgent "attack?"[80] Using these adjective pairs, the author considered it a terrorist attack while labeling the same actions employed against military targets as guerrilla tactics, further illustrating a lack of objectivity and precision in characterizing the conflict. Opposition to French and French-supported governments in Northwest Africa in the same Malian conflict have been described as being motivated by "virulent jihadist ideologies,"[81] ignoring the historical experiences in a region defined by its centuries of foreign colonialism, violence, and economic exploitation. Following in the path of French colonial officers Lyautey and Trinquier, modern counterinsurgents see the "virulent" Northwest African resistance as simply a disease to be eradicated.

Existing theories are problematic from the start if they employ subjective, loaded terms.[82] These terms, such as "terrorist" versus "freedom fighter" act as heuristics fostering cognitive bias and increasing the danger of miscalculation, misrepresentation, and misunderstanding in actual military decision-making because the use of heuristics in the absence of, or willful ignorance toward, the facts has been linked empirically to cognitive distortion.[83] These terms and value-laden descriptions create context from whole cloth, generate artificial patterns, and construct deliberate "us versus them" frameworks that only become more difficult to break as the conflict develops.[84] Like terrorism, "extremism" is a relative term, with both almost always being exonyms. Charles Tilly observed that "terror is a strategy, not a creed," while also noting that the term itself "always refers to someone else's behavior."[85] Framing terrorism as a conflict strategy rather than a specific form of violence, violent though that strategy may be, is perhaps a better use of the term.

Chapter 4. Problems of Scope, Method, Bias, and Character 73

Terror as a strategy comports with the definition of strategy used in game theory, whereby a "strategic move," such as terrorism, conventional ground invasion, or nuclear first strike, all work to influence the opponent's choices in a restrictive manner while also modifying their own expectations of a player's actions.[86] At the same time, the initiator of the strategic move is subject to their own new constraints due to the restrictive effects of moves that cannot be undone, limiting future choices.[87] While religious militancy, communism, or irredentism may characterize some conflicts, they are the *sine qua non* for none of them while even the most brilliant strategy cannot be a synecdoche for sound war theory. The failure of some scholars to acknowledge this and dig deeper is indicative of their own refusal or inability to do so rather than serving as an accurate description of the conflicts they seek to explain.

Problem Four: Character

Much of the current thinking regarding irregular war and its warfare attempts to shield the inadequate treatment of the topic by inventing new terms and concepts to give a patina of scholarship and authority to descriptions of phenomena that these theories and arguments fail to explain. The worried efforts of scholars to use proper verbiage in describing irregular war is merely a distraction, dazzling the reader with "a complex of seriously confused questions by slogans like total war, psychological war, subversive war, insurrectional war, invisible war."[88] This trend is apparent in both international relations and military studies, especially for those attempting to explain irregular wars.[89]

A propensity exists to create so-called generations in war, with direct implications for irregular war studies. This "generationalizing" of war is a fruitless exercise in minutiae achieving few useful results. While some argue for four generations, others see six and still others argue for seven generations.[90] One scholar split Shi'a militia groups neatly yet inaccurately into four generations since 1979, ignoring the intellectual and ideological origins of these groups well before that date, using Western intervention points as the generational phase lines. Indeed, the author claimed that "[the Shi'a's] next step … appears to be politics"[91] revealing simultaneously a failure to grasp how those groups have maintained persistent political objectives within existing state institutions far in advance of any military activity and a purposeful mischaracterization of those groups. Regardless of the arguments, all are futile exercises contributing little to actual theories of war. Chief among these problematic generational trends is the idea of fourth generation war.[92]

The concept of fourth generation war entered Western military vernacular and began circulating following the publication of an influential four-page article at the sunset of the Cold War.[93] This short article had an outsized effect on conventional military thought in the U.S. even into the era of the Global War on Terror well after 2001. Its hypothetical reflection on the future of war after the Soviet collapse along with its conceptual framework have been used to formulate flawed theories of insurgency, unconventional warfare, and terrorism. The idea was, however, neither unique to the West nor new in military circles even if the Western proponents of this theory misunderstood its origins. Russian military theorists began considering a generational concept of war starting with Russian general Vladimir Slipchenko well before 1989. Their generational concept was based on a completely different framework than the four generations proposed in the four-page article and embraced by the raft of supporters that quickly followed.[94]

The Soviet Union invented the concept of the revolution in military affairs in the early 1970s, even if the concept only came to be understood in Western military writing since the early 1990s.[95] The Soviets assessed that the revolution in military affairs began after the Second World War and would continue developing as science and technology advanced.[96] Official Soviet military doctrine reflected the Russian understanding that any revolution in military affairs was "directly linked with radical changes in weapons and military equipment,"[97] that is to say, the revolution would be strictly technical. The Soviets employed this concept in the context of conventional war, in which the Soviets expected a "leap-like transition from conventional to nuclear [weapons],"[98] while deliberately excluding guerrilla tactics and irregular war concepts from this purely technological view of tactical adaptations. The Soviet idea of a revolution in military affairs transformed in meaning and potency while working its way through the U.S. military and policy community as the focus on counterterrorism supplanted containing Communism as a primary national security objective.

Western proponents of the theory defended a concept of four generations of war centering around technology and tactics, something a grand theory of war should seek precisely to avoid.[99] This concept also focused on specific wars as revolutionary moments in themselves, which, in their novelty and intensity however shocking, were still simply a series of combinations of tactics, technology, and violence.[100] Fourth generation war theorists supported massive conventional violence against civilians in irregular wars, voicing their support for the 1982 Hama massacre in Syria, which those theorists dubbed "the Hama model" and suggested should be replicated by Western militaries.[101] Another influential treatise for the

Chapter 4. Problems of Scope, Method, Bias, and Character   75

acolytes of fourth generation warfare argued that the first generation of war covered all human history up to 1815, bundling thousands of years of war into a single generation while privileging a few decades into several discrete groups equal to those millennia of global combat.[102] War studies tend to look to past conflicts from conventional perspectives for answers in the otiose hope that the next war will be some advanced version of the last.[103] The concept of fourth generation warfare is an analytical error of exactly this type.

This flaw in military studies will not be easily overcome.[104] One influential military theorist considered Martin van Creveld's writings on the generational concept of war an "intellectual *tour de force*,"[105] specifically characterizing van Creveld's 1977 work *Supplying War* as "one of those rare volumes that so dominates a subject that anyone attempting to advance knowledge in the field must confirm it."[106] Other prominent military authors have lavished praise on van Creveld's 1991 book, *Transformation of War*, considering it "easily the most important book on war written in the last quarter century."[107] Proponents of fourth generation war are not limited to the U.S. military. Theorists serving in the Indian and Pakistani militaries consider the fourth-generation war model as sufficient to "describe this new mode of insurgency-driven conflict."[108] Thankfully, the concept is not without its detractors, even if those voices form a minority view.

The idea of evolutions, generations, or revolutions in military affairs are not universally accepted.[109] One argument against fourth generation war stands behind Clausewitz in asserting that the nature of war is unchanging, despite its fluid character.[110] In this view, war is not strategy and nothing is really novel about "modern" war despite the advent of technology or firepower, as these are character-affecting and not nature-changing.[111] The flurry of terms like hybrid warfare, asymmetric warfare, gray zone conflict, revolutions in military affairs, lawfare, mosaic warfare, kill webs, and so on simply reveal how challenged contemporary military theorists have been to maneuver outside the comforts of conventional thought.

Some focus on the semantics of terms like "hybrid warfare" rather than attempting to create a framework explaining the phenomenon those faulty terms seek to name. One well-intentioned but unsuccessful attempt to replace "hybrid warfare" with the term "non-linear warfare" resulted in a definition identical to that of attrition, whereby "The overall objective of a state engaging in [non-linear warfare] is the wearing down of the enemy socially, politically, and militarily."[112] This unoriginal definition mirrors more closely the latter stages of the First World War rather than resistance to Israel's invasion of Lebanon in 1982 and it is hardly an archetype of non-linearity or a break from convention. Instead, "hybrid warfare" is just another neologism for technological advancement in the tactical and operational levels of warfare, not unlike other vague terms vacant of substance at the levels of strategy and theory.[113] Like "hybrid warfare," "asymmetric warfare" is another adjectival phrase referring to actions in which opposing forces are unlike one another in whatever way the observer is measuring the contest of wills. "Asymmetric warfare" became part of official U.S. policy with the 1997 National Security Strategy but was dropped in the 2002 version, in part due to its imprecise meaning.[114] Terms like hybrid war, asymmetric warfare, gray zone conflict, and fourth generation war create unnecessary confusion when clarity is needed, and they represent the difficulty American military thinkers have in dealing with novelty, change, and irregularity rather than representing innovations in theory or paradigm shifts in war.[115] These problems with the concepts themselves are not unique to the U.S., largely because of the outsized influence that U.S. military thought has on its allies, partners, and even its adversaries.

There has been no revolution in theory, strategy, or tactics, but rather in technology. The so-called revolutions in military affairs are simply strategic frameworks for employing new technologies rather than true historical changes. Adherents to this misguided set of views fail to furnish explanations of any phenomena that can sufficiently lead to the development of a military theory that is both novel and valid.[116] Despite this, one military theorist closely linked to U.S. military policy formulation confirmed that by 2004, the concept of a revolution in military affairs "had achieved a status close to orthodoxy in the U.S. policy community."[117] That preference did not disappear despite competing theories advanced during the subsequent decades.[118] This confusion of generations, evolutions, and revolutions creates artificial gaps in both scholarly literature and military doctrine where there has long existed a poor understanding of irregular war problems.[119] Instead, a broader, more useful, and realistic theory of irregular war must analyze conflict through a prism of economics, politics, and society while being tempered by the notion that no war in the future will be like any in the past.

The Human Element

Carl von Clausewitz acknowledged in 1832 that "theory must always take into account the human element."[120] It is neither a tactical nor doctrinal revolution in military thought that is required to fight and succeed in irregular wars, because the chief actors in these wars are humans and not hardware. Instead, a theory must be developed to explain the phenomena

Chapter 4. Problems of Scope, Method, Bias, and Character 77

at work in these conflicts, phenomena revolving around the human element. Following this, a suitable grand strategy to achieve a political objective can finally be developed for use in irregular wars. For Clausewitz, the theory of war was framed as the interplay between irrationality, reason, and chance. These three elements were personified as the raw violence of man, the constraining politics of the national state, and the unforeseen effects of interacting forces, natural and human. We have now reached the important question of what factors are at work when structural changes to social orders, sovereign territories, and political systems lead to irregular war, since not all structural changes result in such conflicts.

Theoretical principles are necessary to develop grand strategy from the perspective of irregular war. If wars are most often the result of failed grand strategies as some argue, then states such as the U.S., Israel, and Russia have consistently engaged in failing foreign policies.[121] Irregular war is often simultaneously a cause, condition, and effect of those failures. Its character must therefore be studied further. To do this, we will examine the inherently philosophical conditions of irregular war, including agency, legitimacy, and sovereignty. But first, existing conventional war theory, doctrine, and force design must be examined so as to identify the precise location of their flaws when used in irregular wars so that a more accurate explanation can take their place.

Chapter 5

Conventional War Theory and Regular Wars

> The guerrilla wins if he does not lose; the conventional army loses if it does not win.—HENRY KISSINGER[1]

Some characteristics of conflicts stand out in their expressions of organized violence. These may appear traditional, conventional, and regular, while others may appear novel, unconventional, and irregular. Ultimately, the participants, their objectives, and the structure of the conflict define the character of the war, not the observers of that structure. Similarly, no matter how elegant, lengthy, or forceful a theory of war may be, it may well remain unsuited for explaining conflicts outside its scope. Conventional war theories framing conflict in the context of states suffer from this problem, which leads to failure, frustration, and unnecessary loss of life in operations conducted under their aegis. U.S. military doctrine is the chief example. Fortunately, frameworks better suited for explaining irregular wars are available in the social sciences, and two stand out in particular for this study.

The character of irregular war differs from traditional war in its conduct, the parties to the conflict, and the objectives especially of the participants that are not state militaries. Conventional war theories promote certain primary strategies for engaging in conflict with other states.[2] These strategies are containment, restriction, isolation, balancing, and stirring. The latter, stirring, has direct implications for third-party interventions in irregular wars, as it often employs unconventional warfare, subversion through covert action, and overt military involvement.

Theoretical frameworks from the field of military theory are understandably bogged down in tactical and operational factors such as terrain, time, and technology since the intended consumers of these frameworks are *traditional military forces* engaged in *regular wars*. It is therefore true that industrial potential and population size are useful heuristics for predicting military force composition and strength in conventional contexts but not in irregular wars.³

Law and Doctrine

The basic premise of irregular war remains distinct regardless of which adjectives military theorists use to describe that premise or which concepts are included in the doctrinal publications implementing the policy objectives of those who instruct combatants to fight in these conflicts. The war doctrine of the U.S. military is a powerful example of a military-theoretical framework that is unsuitable for a study of irregular war. Its theoretical assumptions have been felt globally as it frames current approaches to these conflicts. Ironically, most traditional military-theoretical frameworks, including U.S. military doctrine, are ill-equipped for explaining irregular wars despite incessant involvement in these types of conflicts throughout the nation's history. A brief examination of U.S. military doctrine serves as a useful heuristic for understanding why this is the case.

For the U.S. military, *war* is a violent clash of interests, and a *regular war* is "a declared war between the U.S. and any foreign nation or government, or any invasion or predatory incursion [that] is perpetrated, attempted, or threatened against the territory of the U.S. by any foreign nation or government," according to both U.S. law and Joint military doctrine.⁴ This definition describes the adversary in a regular war as another Westphalian state, putting emphasis on state violence and sovereign territory. However, neither the involvement of the state nor the presence of violence alone are sufficient conditions for differentiating between regular and irregular wars, or wars at all.⁵ American law and doctrine therefore narrowly conceive of war, resulting in definitions emphasizing tradition, regularity, and convention. This is totally acceptable for conventional wars between states, but irregular war is no such thing.

U.S. law clearly defines when military force should be used within the 1973 War Powers Resolution. In that statute, *traditional military forces* shall only be used in three cases: "(1) a declaration of war, (2) specific statutory authorization, or (3) a national emergency created by attack upon the United States, its territories or possessions, or its armed forces."⁶ In this sense, Special Operations Forces are also traditional military forces since they fall under Title 10 U.S. Code, even if executing support programs for irregular forces combating terrorism under Title 10, section 127e or supporting Title 50 activities such as those the CIA conducts.⁷ Confusingly, U.S. military doctrine acknowledges that "most U.S. operations since the

11 September 2001 terrorist attacks have been irregular,"[8] while in the same breath rejecting the notion of a difference between regular and irregular wars. The authors of this doctrine therefore followed the same erroneous path of military theorists such as Baron Jomini, Martin van Creveld, and David Kilcullen in misunderstanding what Carl von Clausewitz articulated so well.

American law has done only slightly better than U.S. military doctrine. Following the attacks on 11 September 2001, Congress passed a law known as the Authorization for Use of Military Force.[9] This authorized the President to use traditional military resources, in other words Title 10 forces, against "nations, organizations, or persons" affiliated with the attack, an example of the mismatch between conventional, traditional, and regular state military forces employed against an unconventional, novel, and irregular adversary, keeping in mind the fact that special operations forces are "irregular regulars" bound ultimately by conventional doctrine to conduct traditional U.S. military activities.[10]

Although U.S. military doctrine defines war as "socially sanctioned violence to achieve a political purpose ... an integral aspect of human culture,"[11] the form of combat for which this doctrine is written is strictly regular and traditional. More simply, this doctrine understands war to be "the use of *military* force,"[12] which by the very nature of the belligerents described therein requires regular military forces for its conduct. Conventional military doctrine is written for those forces who will conduct military action for the state. U.S. military doctrine, for example, clearly approaches war from the perspective of a state waging war against another state, despite most U.S. military actions since the founding of the country involving adversaries that are not states.

U.S. military doctrine has yet to reconcile regular and irregular war, instead opting to fit the square peg of irregular war into the round hole of regular war. To do this, a certain logic has entered doctrinal publications to attempt to bridge the gap between the two. For example, the cornerstone document containing Joint U.S. military doctrine used the following roundabout way of integrating non-state actors into its overall framework for explaining war while still adhering to a statute built for states fighting states:

> As the conflicts approach the requirement for the use of force to achieve that nation's interests, military means become predominant and war can result. The emergence of non-state actors has not changed this concept. Non-state actors may not use statecraft as established; however, they do coerce and threaten the diplomatic power of other nations and have used force, terrorism, or support to insurgency to compel a government to act or refrain from acting in a particular situation or manner or to change the government's policies or organization.[13]

This attempted integration of non-state actors into a theoretical framework for war merely conflates the conduct and effects of terrorism, a tactic, with the overall concept of irregular war while reifying the insurgency and alienating the people from the state, ignoring the qualification that terrorism is not a strategy but a tactic.[14]

Contemporary Western doctrine for counterinsurgency and counterterrorism are derived directly from "a selective body of historical work."[15] This body of material is drawn principally from the writings of French and British colonial officers who were known for their violent, racist, and inhumane approaches to suppressing any local resistance to their imperial rule. These approaches do not prioritize objectively understanding the adversary, ignorant to the important admonition in Sun Tzu that "if you know the enemy and know yourself, you need not fear the result of a hundred battles. If you know yourself but not the enemy, for every victory gained you will also suffer a defeat."[16] Contemporary Western military doctrine places the traditional military force at the center of the universe, around which all other actors must revolve.

Current doctrine suffers from three failures in terms of irregular war theory. First, the doctrine fails to adequately explain irregular war. Second, it fails to link causes with effects especially in insurgencies. And finally, the doctrine fails to see beyond the military instrument of national power. These three failures doom the framework into incommensurability with irregular war and serve as a powerful reminder that the synecdoche trap and bias of seeing like a state are the dominant forces in military theory formulation.

Doctrinal Disconnect

The disconnect between the U.S. military's theoretical framework for war and its actual conduct of that war is illustrated in its doctrinal discussion of strategy. Joint U.S. military doctrine gives only two possible strategies in war: annihilation and erosion.[17] Such a narrow view of military strategy places primacy on either conquest or attrition rather than addressing fundamental changes in social order, sovereign territory, or political-economic institutions. Meanwhile, some authors misinterpret Clausewitz, seeing the only strategies in war to consist of some form of total destruction, attrition, or treaty.[18] These strategies share an affinity with the two purposes Niccolò Machiavelli gave for a state to go to war against another state. In his view, "There are two reasons to wage war upon a republic: first, to become its ruler; second, for fear it will take possession of your own."[19] This stands in stark contrast to the objectives for belligerents in irregular wars. In other words, U.S. military doctrine frames military strategy as either attack or defense of the state, yet irregular wars involve the violent transformation of an existing state from beneath the level of regular war characterized by Westphalian combat.

The U.S. military's theoretical framework takes a Westphalian approach to describing war and conducting warfare while using regular war concepts to engage in irregular wars. The resulting attempts to insert non-state actors into the preordained understanding of war in terms of traditional military activities while remaining state-centered has led to a profusion of legal gray areas, questionable intelligence activities, and long-term military failures.[20] It is far easier for states to ignore, justify, or hide their own activities than it is for belligerents who are not states, because the latter lack representation in the international community and are often recognized as criminals, terrorists, and unlawful enemy combatants.[21] This is why the U.S. military artificially categorizes a complex web of problems and solutions into five simple concepts serving policymakers and executive agents with what one observer described as "an icon of strategic assuredness" toward a "magically effective ... all-purpose solution to many irregular enemies" instead of taking the necessary steps toward understanding and then participating in irregular wars.[22] However, these are simply operating concepts that may occur within a campaign, none of which explains or describes concepts of war in general. However, it is organizationally expedient for the U.S. government to separate irregular warfare into the neat categories described in U.S. Joint special operations doctrine from both a budgetary and operational perspective.[23]

Consider the following hypothetical scenarios of how Congress might use those five categories as heuristics for oversight of the Executive. Congress must fund Title 50 national security and Title 10 traditional military activities through defense authorizations and appropriations while also conducting oversight actions such as holding hearings and requesting periodic reports. Congress can use the five irregular warfare categories to craft appropriations more easily for seemingly disparate operations into a single bill, such as section 1207 of the 2006 National Defense Authorization Act that authorized the Department of Defense, mostly via Special Operations Command, to use 100 million dollars annually on "programs that support security ... in countries that are unstable."[24] Congress can then make appropriations for a recurring foreign internal defense mission for a U.S. Navy Special Warfare unit in Yemen, a stability operation at a platform for a Marine Special Operations Team in Niger, an unconventional warfare operation for U.S. Army Special Forces in Mali, a counterterrorism operation for the National Mission Force in Lebanon, and a counterinsurgency operation for a Special Operations Task Force in the

Chapter 5. Conventional War Theory and Regular Wars

Philippines. Meanwhile, the same appropriation can fund U.S. Air Force Special Operations Command to provide forces for air support, combat control, and personnel recovery for all these operations and more. In this scenario the convenience of separating irregular warfare into neat categories allows policymakers and planners to work within authorities, permissions, and organizational relationships to focus the tasked elements to achieve national security objectives. However useful this may be for streamlining the bureaucratic process, this framework is unsuited for developing or employing a theory of irregular war even if the monies are appropriated with the best intentions for engaging in such conflicts.

The use of doctrine in irregular warfare operations as an alternative to understanding or explaining a conflict is also a problem in general with the U.S. approach to security assistance and security cooperation. Security Assistance is a U.S. State Department activity, while Security Cooperation is the preserve of the U.S. Department of Defense. These are mutually supporting activities that can blur in cases such as Foreign Military Sales or International Military Education and Training programs in which the U.S. seeks to build the host nation's security forces in the image of the U.S. with relatively little consideration for culture, norms, and social orders in those countries.[25] For example, American security force assistance programs have imposed American conventional war traditions upon Arab militaries despite the fundamental differences in organizational culture between the sender and the recipient.[26] The effects of this approach are high capability forces with low capacity for employment once the American crutch is removed, as the 2021 withdrawal from Afghanistan and 2011 withdrawal from Iraq both proved.

Another major problem with this approach is that it addresses only warfare, not war. Indeed, U.S. military doctrine claims that the division in war between traditional and irregular occurs only at the level of *warfare*, explicitly rejecting the concept of irregular *war*, the conduct of which is anti–Westphalian.[27] Rather than pursue theories of irregular and regular war, the U.S. military attempted to create a unified framework for understanding conflict beyond these using the concept of the "competition continuum." This concept, according to U.S. military doctrine, sees the world as "a competitive operational environment" characterized by "enduring competition conducted through a mixture of cooperation, competition below armed conflict, and armed conflict."[28] Similarly, influential scholars focus on the *how* but seldom the *why* of war, dismissing the necessity of first considering causes.[29] These authors focus on "a continuous spectrum of military methods," eloquently defending and perpetuating the flawed competition continuum and spectrum of conflict model upon which the U.S. Department of Defense relies to frame war generally.[30]

This hollow idea creates a false impression that all societies are either at war or in competition, failing to account for the internal systems of states, the interaction between states and non-state groups seeking sovereignty, and the mechanisms behind the balancing among groups of states against others. Such an approach employs the "black box" concept of the state found in Realist theories of international political systems and fails to meaningfully engage with these theories in a way that would potentially change the structure of current military theories to address the problems various Realists identify in their own theoretical frameworks.[31]

What is traditional and what is not traditional is a completely subjective judgment. This is true whether discussing war or warfare. Here the bias in U.S. military doctrine is highlighted, whereby doctrine succumbs to the synecdoche trap, seeing like a state, confusing scope, and failing to grasp major causal factors in irregular war. As one social scientist put it, synecdoche and conflation are "good for academic careers but not intellectual progress."[32] This problem is not new. Writing in 1790, British intellectual Edmund Burke was repulsed by the idea that theory might invade civil society, fearing "the moral and political evils that follow upon the intrusion of theory into political practice."[33] Unfortunately, the authors of U.S. military doctrine continue siding inadvertently with Burke.

Wars, Conventional and Regular

There are numerous theoretical frameworks for conventional approaches to war, some more useful than others, and none developed specifically to address irregular war. The dominant but separate fields treating irregular war are comparative history, political science, and political sociology.[34] Most theoretical frameworks lack interdisciplinary breadth for explaining the phenomenon of irregular war. Frameworks in political sociology come close, but most approaches tend to favor the state, if acknowledging any other legitimate actor at all. The Westphalian state system logically dominates theoretical frameworks in political science since this is currently the dominant form of sovereign political organization. Likewise, state-centered frameworks favor political institutions and the state in their theories while neglecting individuals and groups.[35] Ironically, the frameworks found in political sociology and political science are far superior to the military-theoretical frameworks underpinning Western military doctrine. Still, all these frameworks do little to explain irregular wars despite this form of conflict being inextricably linked to the very existence of such states. Thus, such theoretical frameworks are a useful starting point even if they fail to sufficiently link the individual and the state to explain how this relationship contributes to the dysfunctions leading to irregular wars.

U.S. military doctrine uses a theoretical framework that is disconnected from the reality of irregular war. A theoretical framework for irregular war must instead transcend tactics, surpass operations, and understand all belligerents equally to be broadly applicable and internally consistent. Irregular war is a complex phenomenon involving the culmination of certain tensions between people and states resulting in organized violence. The conditions leading to that culminating point emerge from the unresolved antagonisms arising between a dysfunctional sovereign interacting with the social order, sovereign territory, and political-economic institutions of the state. The elements of irregular war include people, politics, and propaganda, while conventional forces and third parties also play a major role. The irregular war once begun consists of a violent dialectic between people and states seeking to resolve their contradictory objectives through means of force. The outcomes of irregular wars extend far beyond the original objectives of any of the participants.

Chapter 6

Conventional Forces in Irregular Wars

If instead of sending 30,000 soldiers to Iran, you had sent 100 teachers!—Qazi Mohammad[1]

A portable field battery in an anonymous Algiers villa, wired with metal clamps to the nipples or testicles; a roomful of French paratroopers and cigarette smoke; constant questions in a flat, bored voice. Tell us what you know, and the pain will stop. Tell us what you know, and the pain will stop. Then the electricity, the white light behind the eyes and the spasms. Algerian independence fighters who survived the torture called it *la gégène*. The French army called it counter-revolution.—Christopher Othen[2]

Whoever fights with monsters should make sure not become a monster in the process. And if you gaze long into the abyss, the abyss also will gaze into you.—Friedrich Nietzsche[3]

Modern conventional forces are manned, trained, and equipped in the image of the accumulated influences of certain impactful historical moments defining those organizations and the states employing them. Chief among these impactful moments was the fundamental change in the character of conventional war during the European campaigns of 1793, in which France employed an innovative form of warfare against Austria and Prussia that washed over their rigid, orderly system like a flood.[4] This new French warfare induced rapid and sweeping changes to the European way of war, changes which nineteenth-century French military theorist Antoine-Henri Jomini viewed as "the great problem solved by 1800."[5] The essential transformation was from a war of *position* to a war of *maneuver* using combined arms. The current organizational structure of virtually every state military has its origins after Napoleon's campaign in Italy in 1800.[6] Since the Seven Years War (1756–63), combined arms consisted of infantry, cavalry, and artillery. In the First World War, armor was added. In the Second World War, the physical domains of war expanded to land, sea, and air all under the supreme command of an individual military leader, requiring massive doctrinal and tactical changes toward new conventions almost universally adopted by all state belligerents through the course of the war.

Punctuated equilibrium crystallizes the habits of these tradition-bound and doctrine-led conventional military bureaucracies over long periods.[7] These habits wield great influence over military decision-making regardless of the wisdom of their applicability in a particular conflict. All this is not to say that conventional forces have no place; indeed, they are perfectly suited for conventional wars. Irregular war happens to be something different. It is therefore prudent to analyze the development of the modern conventional military force to reveal the causes of its habitual deficiencies in irregular wars.

Modern conventional forces are an extension of the Westphalian system of state sovereignty born in that germ of modern conventional wars and of states themselves, the Thirty Years War (1618–48). The state's authority, influence, and ideology define the form of the conventional force while the force's character is shaped through functional and social imperatives.[8] Conventional forces therefore exist to affirm, transform, create, or destroy a state through warfare against other states.[9] The affirmative function of their militaries often works with the mere threat of force rather than its much less common use. However, new states are formed from the European model of state formation quite rarely: the transformative, creative, and destructive functions of conventional forces often sit as capacity at rest for long periods of time.[10] Despite militaries using only a minority of their functional purpose at any given time, the conventional military theories derived from these capabilities usually find their utility outside the parent state rather than within it. This reality casts the broad shadow of colonialism, world wars, and globalization. Further, the Realist concept of the security dilemma leads states to build conventional capabilities to counter those of other states because states can never rule out conventional attack in the Westphalian system.[11]

Combatting Bandits in Small Wars

By design, conventional forces provide security relative only to other states. This is not a novel concept. In his 1531 essay on conspiracies, Niccolò Machiavelli assumed it was common knowledge that heads of state lost both their offices and their lives more frequently to internal conspiracies than from wars between states. Realist explanations of war do not explain war within states because theorists in international relations and political science separate civil from interstate conflict along an anarchy-hierarchy dichotomy following from the differences in academic focus between those fields rather than from ground truth.[12] As with the Realist school, there is a deficiency in U.S. military thought regarding irregular war, as theorists and practitioners artificially separate insurgencies from civil wars and revolutions.[13] The first modern conventional force, the Royal Prussian Army, was forged in just such a crucible of state warring against state.[14]

Prussia's spectacular loss to Napoleon at Jena-Auerstedt in October 1806 instigated its transformation into the first modern conventional force placing a premium on education, technical proficiency, valor, and perception and away from a common European model little changed since Late Antiquity.[15] Prussia supplanted Napoleonic France to become "the most famous, prototypical military organization of the Eurocentric world" from 1806 to 1871, becoming "a classical example of an army organization radically repressing the very thought of partisanship."[16] Prussia inevitably became "not a country with an army, but an army with a country."[17] Although Prussia produced von Clausewitz, von Moltke, von Gneisenau, von Scharnhorst, and other important military theorists, contemporary Western powers continued relying on Napoleonic concepts to shape their doctrine, tradition, and concepts of employment.

Nineteenth-century American military education was based on a worshipful admiration of French military science, which emphasized scientific engineering concepts, the construction and destruction of fortifications, topographic study, maneuver warfare, the science of ballistics, and the use of artillery pieces.[18] Influential American military theorist Dennis Hart Mahan founded a Napoleon seminar at West Point, while his son and prominent naval theorist Alfred Thayer Mahan attended French military academies after graduating from West Point. The elder Mahan was a chief architect of the importation of European military thought, particularly from France, into the U.S. military officer training curriculum.[19] The French geometric approach to war was made famous with the writings of Antoine-Henri Jomini and this trend ended up leading American theorists away from the actual conduct of war emphasized in Prussian schools and battlefields through 1871. However, a transformation occurred in the waning years of the nineteenth century and by the outbreak of the First World War, American military theories of war and war's relationship to the state were firmly back within the classic Prussian ambit framed most clearly by Carl von Clausewitz.[20]

Whether originating in Prussian or Napoleonic traditions, modern conventional forces such as the U.S. military are relatively inflexible and risk averse. American General Wesley Clark, commander of Operation Allied Force in Kosovo (1999), lamented that "the military's innate conservatism" often hamstrings its own actualization.[21] One study broke this conservatism problem down into thirteen characteristics that define the American way of war: it is apolitical, astrategic, ahistorical, optimistic, culturally challenged, technologically dependent, focused on firepower, large-scale, aggressive-offensive, profoundly regular, impatient, logistically excellent, and highly sensitive to casualties.[22] Most of these characteristics feature in the 1984 Weinberger Doctrine and its application in planning and executing Operations Desert Shield and Desert Storm in 1991. The force employment concept in the subsequent Powell Doctrine complimented those elements of the American way of war that preferred short, decisive wars with overwhelming odds of victory.[23]

Despite paradigm shifts in technology, conventional military forces use concepts that are hundreds or even thousands of years old. For example, aerial reconnaissance has come a long way from "a frail basket floating in mid-air," as Antoine-Henri Jomini affectionately described its first military use during the Battle of Fleurus in 1794.[24] Long-range communications have similarly improved since the telegraph was first used in war at the Battle of Ratisbon in 1809, in which Napoleon was notified in Paris by optical telegraph about the Austrian advance across the River Inn, a transmission that took 24 hours to pass over 700 miles.[25] But the organization of forces and their concepts of employment have a much deeper history.

The concepts of a national military, maneuver warfare, concentration of firepower on a decisive point, defense in depth, branch planning, and others arose in a specific conventional military context influenced more by tradition than expedience. The modern concept of the national military arose under Oliver Cromwell in the 1650s in contrast to forces traditionally tied to specific garrisons.[26] Maneuver warfare developed in opposition to positional warfare, and the German theorists who first developed the concept believed it could be used in any context because all war should terminate with confronting and destroying the enemy in open battle, which was the only form of war German thinkers were willing to acknowledge at the time.[27] Concentration of firepower on a *Schwerpunkt* (center of gravity/decisive point) became a tactical crutch on which Napoleon rested after 1807 as he began eschewing the creative strategies he employed in the earlier battles at Jena-Auerstedt, Ulm, Austerlitz, and Marengo.[28] The concept of the enemy center of gravity (from Italian: *centra gravitatis*) was first directed at massed military formations in the theory of defense of theaters of war found in the writings of Carl von Clausewitz.[29] This was not directed at civilians, infrastructure, or supply systems, contrary to its current application in conventional military doctrine. Much earlier, Sun Tzu illustrated the concept of massing forces against enemy weak points in his discussion of energy, likening the concept to "a grindstone smashed against an egg."[30] However, Mao Zedong's observation that "the war of position and the war of maneuver differ fundamentally from guerrilla warfare" cannot be ignored.[31] There is no egg to smash in irregular wars.

The ideas of defense in depth and the military planning process have similar histories. Defense in depth developed as a NATO strategy to impede a sub-nuclear Soviet invasion of West Germany, itself a derivative of an older hedgehog defense strategy designed to slow offensive tempo in the same territory.[32] In the 1920s, German officer Joachim von Stülpnagel promoted the concept of *Volkskrieg* (people's war) as an element of defense-in-depth around Germany's borders.[33] In terms of military planning, the use of branches and sequels was an artifact of Bourcet's 1775 treatise *Principes de la guerre de montagnes* [Principles of Mountain Warfare], which gained wide use during the Napoleonic Wars.

More than these concepts of employment, common orders of battle for modern military forces are a distinctive product of European history in general and Westphalian wars in particular. This is true whether fighting on land or at sea. The Thirty Years War (1618–48) produced the 550-man battalion divided into companies and platoons along with the officer and non-commissioned officer leadership system needed to field this organizational structure in combat formations, a process that had actually begun in 1611 a few years prior to the war.[34] The battalion had already become the primary maneuver element before 1611 in Sweden under Gustavus Adolphus, while standardization of unit formations employing projectile-firing weapons had emerged earlier in the Netherlands by 1599.[35]

These structures were all modified revivals of the Roman *manipulus* developed in the fourth century BCE. The maniple consisted of three ranks of 40 soldiers each, for a total of 120 men, about a modern company-sized element. The Marian reforms in the second century BCE focused on the *cohors* as the primary fighting unit, comprised of around 480 men, roughly equivalent to a modern battalion-sized element. Ten cohorts comprised a legion with 5,000 men, equivalent to a modern regiment or brigade.[36] Similarly, naval orders of battle remained virtually unchanged from 1660 to 1918.

Modern military uniforms also arose with the French in the Thirty Years War (1618–48), which "made soldiers look like soldiers as we know them in the twentieth century."[37] Western clothing and visible modernization initiatives were not restricted to the European powers, as new fashions arrived in Turkey and Iran in the 1920s through the vector of European military uniforms after the First World War. This clothing

Chapter 6. Conventional Forces in Irregular Wars 91

became a target during subsequent revolutionary and opposition movements in those states, in which one such movement described the practice as "dressing like infidels."[38] Despite this opposition, the militaries in every state, including those in Iran and North Korea, use a military service uniform based on styles first developed in a war that ended four centuries ago in Europe.

Naval traditions are no exception to the conventions emanating from the European continent since the Peace of Westphalia. The distribution of combined arms among vessels grouped into squadrons featured prominently in the First Anglo-Dutch War (1652–54), with a flagship and naval group command structure similar to the structure in use in 1890 when Alfred Thayer Mahan lauded that concept in his treatise on sea power. This rigid, predictable, and unimaginative conventional formation is the maritime equivalent of the land warfare Antoine-Henri Jomini promoted and that persists as a popular topic of study among military theorists and practitioners. Current U.S. Navy carrier strike group formation concepts remain remarkably consistent with those used in the seventeenth century, despite paradigm shifts in technology since that time, including submarines, aircraft integrated with carrier strike groups, nuclear reactors, and ballistic missiles. Again, this formation concept is used worldwide, even among states such as Iran, China, and North Korea.

The propensity for unimaginative adherence to convention may explain why the U.S. did not produce a national military strategy until 1986. The U.S. also lacked a concept of the operational level of war until 1981, despite the German *blitzkrieg* offensive in the Second World War demonstrating the need for it during Operations Yellow and Red.[39] In contrast, China's first national military strategy was released in 1956, only seven years after its founding, and focused on a combined arms operation to repel an American invasion that the Chinese Communist Party feared was inevitable.[40] Interestingly, Mao Zedong was removed from military command in 1932 because the Red Army initially rejected his ideas of "luring in deep" and "active defense."[41]

It was only after a series of conventional Red Army defeats at the hands of the Kuomintang and the Japanese that Mao was able to return and promote his idea of *yu chi chan* for which he is now well known. *Yu chi chan/Youji Zhan*, often translated to English as guerrilla warfare, more accurately translates as "small-scale harassment and sabotage operations behind enemy lines."[42] Although the modern Chinese state was born out of an irregular war ending in 1939, its first military strategy focused on conventional defense in depth against a state, a telling example of the pervasive application of Westphalian conventional force concepts worldwide.

How did so many states come to adopt a conventional approach mismatched to irregular wars? In May 1890, Prussian field marshal Helmuthe von Moltke stood before the Reichstag and perspicaciously argued that "the age of *Kabinettskrieg* is behind us—now we only have people's war."[43] *Kabinettskrieg* describes "short, limited conflicts fought largely for territorial objectives which were easily satisfied"[44] between 1648 and 1789, though some argue that the Crimean War (1853–56) was the last *Kabinettskrieg* and others consider it the first modern war.[45] Regardless of the semantics involved, Carl von Clausewitz warned "Woe to the *Kabinett* which, with a shilly-shally policy, and a routine-ridden military system, meets with an adversary who ... knows no other law than that of his intrinsic force."[46] The Prussians had much to fear and to appreciate from irregular forces throughout the nineteenth century, just as other conventional forces did.

Irregular forces have featured in most conventional wars since the Peace of Westphalia. Early examples of irregulars converging with conventional forces include *chaussers* (hunters), light troops, and partisans, all common in the eighteenth and nineteenth centuries in Europe. Prussia employed the *Landwehr*, a people's army employed as an auxiliary force.[47] The German *bandenkrieg* and the Austrian *jagdkampf* were roughly equivalent to light infantry, commandos, and partisans, all of whom operated on behalf of a state, and were first employed in the Thirty Years War (1618–48).[48] The partisan played an important role in the earliest Westphalian conflicts, even during the Thirty Years War itself. On 24 November 1643, Hapsburg irregulars under Frantz von Mercy infiltrated an enemy encampment and then killed over 4,000 of the Holy Roman Emperor's forces under Josias Rantzau at the Battle of Tuttlingen.[49] In the eighteenth century, partisans operated between Denain and Bouchain in 1712, while the Hapsburg partisans supported conventional forces against Prussian supply convoys in 1758.

In a perhaps better-known example, Britain enlisted men to fight against what the British government considered "a rebellion or disturbance" in the American colonies in the 1770s against George Washington's "undisciplined bandits."[50] One of these men was Andreas Emmerich, a Hessian mercenary in service to the British. Emmerich described himself as a partisan fighting on behalf of the British earlier in the Seven Years War (1756–63). He then led a group of *chasseurs* against the colonists during the American Revolution (1765–83). Emmerich described these *chasseurs* as "a certain type of light infantry troop,"[51] carrying the connotation of hunters from the original French. Emmerich organized his *chasseurs* into battalions supported by grenadiers, all operating together in teams of 150 and with no more than 1,700 men in his total force. This was an important contrast to the organizational structure in the conventional

Chapter 6. Conventional Forces in Irregular Wars 93

British main fighting units at the time, which were all regiments led by colonels. Emmerich saw his *chasseurs* as partisans, using a common definition of that term popularized in *Simes Military Medley*, written as a pamphlet by Thomas Simes in 1767.[52] Emmerich also based his interpretation of *chasseurs* and partisans on the example set by the Hanoverian mercenary Wilhelm von Freytag and his *Jaeger Korps* (scout corps), who also fought for the British in the Seven Years War (1756–63). Emmerich's *chasseurs* performed some tasks that are functionally similar to the core activities of modern special operations forces, including special reconnaissance and direct action.[53]

The aforementioned terms for irregular forces in the eighteenth and nineteenth century took on a more general description as the twentieth century unfolded: they were *guerrillas* fighting *guerrilla warfare*. As is well known, the origin of the modern term "guerrilla warfare" (from Spanish *guerra de guerrillas*) was born in the successful combined Spanish-Portuguese resistance under British command against Napoleon's main force in the Spanish Peninsular War (1808–14).[54] Britain's Duke of Wellington, Arthur Wellesley, resorted to working alongside irregular forces during the Peninsular War in part because he could not afford to have France's *Grande Armée* of 230,000 to 320,000 men annihilate Britain's only field army, which was deployed in its entirety to Spain for this conflict.[55] The Spanish guerrillas provided 36,500 to 50,000 additional men, in addition to the Portuguese and Spanish regulars fighting under Wellington's command.[56] Although the modern term *guerrilla* arose from the Peninsular War, it was the German-speaking states who both employed and encountered irregular forces most in Europe during the eighteenth and nineteenth centuries.

In the conventional space, the U.S. Marine Corps "Combat Hunter" program was a direct descendent of the eighteenth-century French *chasseurs*, borrowing also from the German *Jaegerkommando* (rangers/scouts) concept. The *Jaegerkommando* was developed in a futile attempt to defeat the Soviet partisans up to 1944.[57] Charles Thayer, writing on guerrilla warfare during the Vietnam War, observed that "these [Germans] soon learned that stalking the most elusive old stag was far different from pursuing a wily peasant armed with a rifle."[58] Thayer also pointed out that the Germans surrendered a year after they implemented their own combat hunter project while the Soviets continued on, stronger than before the *Jaegerkommando* was used. Thayer considered the German *Bandenbekämpfung* manual to be "singularly devoid of imagination" and featuring "tactics familiar to every German hunter of hares or partridges" and unsuited for application in irregular wars.[59] In his reflection on this type of mismatch, political scientist Samuel Huntington observed that "history is valuable to the military man only when it is used to develop principles which may be capable of future applications,"⁶⁰ and thus far it would seem the value of history for understanding irregular wars has yet to be realized.

It follows that most modern conceptions of partisans, irregulars, and commando activities are derived from Germanic experience. The first European manual on guerrilla warfare was published in 1840 in Austria and it was used rather successfully in their campaign against partisans in Italy in 1848.⁶¹ It would be a century after that first 1840 manual's publication that the German High Command would produce an update with the 1944 pamphlet *Bandenbekämpfung* (Combating Bandits), despite constant engagement in guerrilla and counter-guerrilla activity throughout the Second World War in nearly every area of operations German troops were engaged up to that point.⁶² German soldier Arthur Ehrhardt had already published *Kleinkrieg* (Small Wars) in 1935, which discussed partisans as military actors, *Volkskrieg* (people's war) as a political activity, and how to counter these forces. Although the Wehrmacht ignored his work initially, the U.S. Army Command and Staff College did not. Their first English translation came almost immediately within the year it was published in German.⁶³ Another German officer, Hans von Seeckt, developed *Aufstragstaktik* (mission command/mission-type orders) in the 1920s, introducing the modern understanding of mechanized maneuver warfare and combined arms underpinning conventional military theories of employment and the "purpose, method, end state" or "task and purpose" model used in the U.S. military.⁶⁴ By 1949, the U.S. Army had ordered translations of 721 strategic and historical studies written by Nazi officers during the Second World War, including 21 works dealing directly with insurgency and 44 relating to partisan warfare.⁶⁵ The U.S. Army began using these newfound German prescriptions for countering insurgency and conducting unconventional warfare immediately in the Korean War.⁶⁶

German military doctrine regarding insurgency and counterinsurgency evolved over time following publication of the *Bandenbekämpfung* manual in 1944. The German Army's current manual, *Land Forces in Counterinsurgency*, defines insurgency as "a process of destabilization caused by political, economic, and/or civil grievances, which affects both the effectiveness and legitimacy of the governmental system," while it defines counterinsurgency as "establishing security and state order in crisis areas."⁶⁷ In contrast, the original entry for counterinsurgency found in the 1944 *Bandenbekämpfung* defined it as civic action and counter-guerrilla operations. The U.S. military's counterinsurgency doctrine has also changed over the years, yet it retains the elements from the German understanding of guerrilla warfare developed from 1808 to 1944. In 1951, the U.S. Army published its first field manual on special operations, which drew directly from the

Chapter 6. Conventional Forces in Irregular Wars 95

German *Bandenbekämpfung* manual.[68] The definition of unconventional warfare has changed considerably since its first mention in the U.S. Army's 1951 Field Manual 31–50, but still reveals its connection to German doctrine. The current doctrinal definition the U.S. military uses for unconventional warfare remains focused on supporting a resistance movement against an existing government while also emphasizing escape and evasion through denied or non-permissive areas, all borrowed from the original 1944 German pamphlet.[69] Additionally, von Bülow's concept of lines of operation first introduced in 1799 became a central element of operational design and counterinsurgency operations in the U.S. military, including U.S. Special Operations Forces.

In contrast to the U.S., Britain published its first doctrine on special operations and irregular war in 1939 in a series of three pamphlets known as G.S.(R) 1 through 3. These pamphlets, produced out of the Special Operations Executive, dealt with leading resistance forces, conducting guerrilla warfare, and conducting sabotage activities, respectively. The U.S. did borrow some ideas from the British, however. The origin of the U.S. Special Operations Forces concepts of "operations, actions, and activities" derived from the Special Operations Executive G.S.(R) pamphlet no. 1. This pamphlet defined *operations* as conducted by a military organization under a campaign, *actions* as those conducted by a band under a leader in tactical actions, and *activities* as those conducted by an individual including activities for a special purpose, such as sabotage or sniping. In general, the British emphasized unconventional warfare and subversion in contrast to the U.S., which focused instead on the French and German models of counterinsurgency and stability operations.

Counterinsurgency and stability operations are campaign strategies conventional and security forces may conduct to provide and maintain an alternative sociopolitical system to the one promoted or sustained by an insurgency. Current U.S. counterinsurgency doctrine is enshrined in the U.S. Army Field Manual on counterinsurgency.[70] This manual is mostly an updated version of the U.S. Marine Corps colonial policing guide for Latin America, the *Small Wars Manual*, published in 1940. One supporter of the 2006 U.S. counterinsurgency manual considered it "the most influential official publication on [guerrilla warfare]" in English since the 1940 *Small Wars Manual* and its 1896 antecedent by Charles Callwell, also titled *Small Wars*.[71] One of the field manual's co-authors considered the manual to be "paradigm shattering," though they later admitted the manual was actually based on French colonial experiences in Algeria along with "British colonial teachings and U.S. Marine Corps code of conduct for occupying Latin American nations," the latter a reference again to the 1940 U.S. Marine Corps *Small Wars Manual*.[72] The counterinsurgency manual also drew on the ideas of divide-and-rule proponent T.E. Lawrence, along with two French officers who participated in the failed war to suppress the insurgency in Algeria. This narrowed the utility of the field manual to essentially a military policing guide for a colonial occupying power, an unfortunate anachronism.[73] The 1940 *Small Wars Manual* was itself drawn mainly from Charles Callwell's 1896 work on the same subject, which was principally written for colonial forces at the height of post–Berlin Conference conflicts at the close of the nineteenth century. Problems with Callwell's 1896 manual and the 1940 rehashing of the same still linger on across the pages of the 2006 field manual.

In any event, the U.S. has long struggled with how to label irregular wars. In 2008, the Department of Defense debated whether to drop the term "irregular warfare" in favor of "cooperative security" and "hybrid warfare" to describe counterinsurgency, foreign internal defense, stability operations, and security cooperation to defend against what then–Defense Secretary Robert Gates considered "irregular threats" along the "spectrum of conflict."[74] The U.S. then applied its untested and ineffective counterinsurgency doctrine in Afghanistan, despite its initial development for Iraq, in a manner not unlike that of France in which "the school of Indo-China arrived in Algeria" fresh from defeat at Dien Bien Phu in May 1954.[75]

The U.S. was not the first to face the mismatch between conventional forces and irregular war in the rapidly changing environment after the Allied victories in 1945. France's first foray employing modern conventional forces in an irregular war context after the Second World War took place simultaneously in North Africa and Southeast Asia. France used indigenous *commandos noir* (black commandos) and *harkis* (mover/war party member), both based around a light infantry model before and during the war in Algeria in the 1950s.[76] The French concept of *ratissage* (raking-over) was a brutal counterinsurgency method developed and employed in various wars against insurgencies in Morocco, Algeria, and Indochina.[77] This method inspired David Galula, Robert Taber, and Roger Trinquier, all of whom subsequently informed the counterinsurgency doctrine employed in the U.S. military up to the current counterinsurgency field manual, as already noted. However, this was not the beginning of German or French influence on American conventional forces in irregular war.

History is rich with lessons for conventional forces to apply in irregular wars, but these are lessons lost on organizations resisting innovation, adaptation, and change.[78] French, British, American, Israeli, Russian, and other examples indicate conventional forces perform poorly in irregular wars despite their relative or absolute superiority in those conflicts in terms of technology, manpower, and resources. In a broader

Chapter 6. Conventional Forces in Irregular Wars 97

sense, states fail in irregular wars for more nuanced reasons than military culture, bureaucracy, and the inability to adapt. There is a problem of "military myopia" in major combat operations, whereby some states, especially Western democracies, cannot see beyond the short-term plans of a few years ahead due to the nature of election cycles and other domestic pressures.[79] The U.S. failed its counterinsurgency operation in Vietnam because democracies tend to shield median voters from the costs of military conflicts to maintain favor in domestic politics, and a comprehensive counterinsurgency operation imposes significant costs on those voters. The fallacies of recency and proximity govern public opinion on war just as they do with any other policy issue.

Numerous alternative approaches for conventional forces in irregular wars have been proposed. One potential approach employs the "police primacy principle," in which police rather than the military should be the primary focus in a counterinsurgency operation.[80] Other approaches include society-wide reforms, counterpropaganda, and decolonization as actions a state may take to prevent, slow, or neuter an insurgency.[81] Counterinsurgency operations predicated on military force risk devolving into "an upward spiral of brutality,"[82] in which violence alone demands an escalatory response. Carl von Clausewitz articulated those elements of unmitigated escalation in his discussion of the tempering effects of subordinating irrational violence beneath rational policy within the framework of state-centered war strategies.

However insightful these arguments may be, often ignored in these discussions is the importance of changes in social orders, sovereign territory, or political-economic institutions that instigate the development of insurgencies in the first place. Some observers miss these elements entirely or view causal chains backwards. For example, one study argued that "intense grievances *are produced by* civil war—indeed, this is often a central objective of rebel strategy."[83] In reality, grievances come before insurgencies; the insurgency is not created *ex nihilo*. Here effects are confused with causes to disastrous effect.

For various reasons, states engaging in these conflicts are often limited to specific strategic frameworks for interacting with an insurgency. States may employ four strategies in their interactions with irregular forces: incorporation, containment, collusion, and suppression.[84] While most social sciences literature focuses on collusion, most military theory focuses instead on suppression and containment.[85] Strategies of suppression and containment feature prominently in U.S. doctrine and the literature closely associated with the authorship of that doctrine.[86] One exception is the positive emphasis on collusion in unconventional warfare in the U.S. military's doctrine.[87] Incorporation tends to occur outside conventional military thought, as it emphasizes political and diplomatic efforts.[88] The examples of France, Britain, the U.S., Israel, Russia, and several others indicate that states exceedingly favor suppression above all else and at great cost.

A French Vision of Disaster

The French military experience shaped much of the modern understanding of counterinsurgency operations and state-centered approaches to irregular wars.[89] France encountered irregular wars and insurgencies soon after the French Revolution. France then spent the rest of the nineteenth and early twentieth centuries developing and perfecting their novel strategies for confronting irregulars such as winning "hearts and minds," the conduct of so-called native politics, violent *ratissage* (raking-over) of entire populated areas without regard for which individuals were actually a threat, forced *quadrillage* (gridding) of population centers, and eradicating the *tache d'huile* (oil spot) of insurgents in the civilian population. These strategies were shown to be largely ineffective early on, yet their use persisted well into the next century.

Antoine-Henri Jomini experienced the Spanish *guerrilleros* (small warriors) firsthand with not a little frustration in 1809. Upon pushing his massive army into Spain, his entire supply train was abducted in the middle of the night by "the peasants, led by their monks and priests."[90] By February 1810, France had marshaled 300,000 soldiers against the 50,000 Wellington brought to Spain. Most of the latter force was composed of British-trained Portuguese irregular forces who joined the indigenous *guerrilleros* to wear down the *Grande Armée*. Soon after, France began employing a "native politics" strategy in the conquest of Algeria, Madagascar, and Morocco from 1830 to 1912. This strategy combined elements of winning hearts and minds and its companion *tache d'huile* (oil spot) technique.[91] The "hearts and minds" concept matured in 1895 under French colonial official Joseph Gallieni in the ruthless French adventure in Madagascar. Charles Callwell, author of the original 1896 treatise on small wars, considered that French expedition to be "desultory warfare to be avoided," ending in "little better than a failure."[92] The "hearts and minds" concept was used to make otherwise unpalatable military activity digestible in domestic press and the political milieu more than to engender support of French conventional forces among the indigenous population. One historian of the French conquest of Morocco observed that "if 'hearts and minds' failed as a military doctrine it was brilliantly accepted as a piece of colonialist propaganda."[93] Such colonialist propaganda was necessary,

Chapter 6. Conventional Forces in Irregular Wars 99

given the brutality of French methods in conducting counterinsurgency operations in their colonies, along with their failures.

France dragged irregular war failures forward into the twentieth century as their military filled the vacuum left behind by the withering Ottoman Empire in the Middle East. French General Maurice Serrail decided to indiscriminately bombard Damascus during the Syrian revolt against French rule in 1925, which served to "appall world opinion and galvanize Arab dissidents."[94] The Druze Revolt in Syria and Lebanon (1925–27) that followed shocked the French public with "a vision of disaster,"[95] yet the military campaign persisted with a strategy of suppression developed during the Spanish adventure more than a century earlier in the Napoleonic Wars. In the end, the Druze captured around 2,000 rifles and automatic weapons along with ammunition after killing 600 French soldiers out of the original expedition of 3,000 men. Two decades later in Indochina, France attempted to capture Ho Chi Minh and his staff for five months in 1947 but were met with repeated failures due in part to a population hostile to French rule. France experienced the same challenges in leadership targeting operations a decade later in Algeria.[96]

France abandoned any trace of the comparatively soft approach of native politics and winning "hearts and minds" in favor of direct action when French forces moved from Indochina to Algeria in 1954. The French developed the so-called "smallpox chart" in Algeria to map FLN activity and cells, with its name giving some indication of France's attitude toward the Algerian desire for independence. For France, the cure to the Algerian smallpox, which was the idea of independence from French colonial rule, could only be the purifying effects of *ratissage* (raking-over) and overwhelming numerical superiority.[97] The exact numerical advantage varies based on the source consulted. According to one widely accepted figure, the ratio of forces during the Algerian war for independence was about 20 to 1, with 150,000 French forces against 7,000 FLN fighters.[98] The number of FLN "fighters" cited in these and other figures are likely inflated, because Algerians in support, auxiliary, and sympathetic roles to the FLN were not actually bearing arms in the conflict but may have been counted anyway. Innocent Algerians captured and tortured by French *paras* were of course also counted as belligerents, since it would be embarrassing to acknowledge French mistakes during the war. Regardless of the numbers, the French method favored quantity over quality while emphasizing shock value over measurable effects. Additionally, the number of French forces in all counts excluded the French village self-defense units in all areas, a physical fence built by the French, and a French naval blockade. The war also produced several memoirs that continue enjoying doting attention in Western military circles for their insights on irregular wars in general and counterinsurgency operations in particular, even though France lost the war due in no small part to its reliance on the methods celebrated in those writings.

Of these authors, Roger Trinquier was a major contributor to U.S. counterinsurgency thinking. Trinquier led the *Dispositif de Protection Urbaine* (Urban Protection System) in Algiers, employing practices that one historian of the war described as having "sinister undertones that also could not help but recall French experiences under [Nazi Germany]."[99] Roger Trinquier was long considered to be "one of the leading Indo-China hands" and "*The* expert on Far Eastern subversive warfare."[100] In his own memoir of the war, Trinquier lauded the failed French counterinsurgency strategy in Indochina as a success in his own book following the conflict, not unlike U.S. Army Generals David Petraeus, Stanley McChrystal, and William Westmoreland following their leadership during the irregular wars in Iraq, Afghanistan, and Vietnam, respectively.[101] Far from an expert in guerrilla warfare, Trinquier instead subjected Algiers to conditions "unpleasantly reminiscent of the Third Reich."[102] However, this strategy was not born in Algeria. A few years earlier, France had perfected the *quadrillage* (gridding) method of forced relocation of civilians into concentration camps in Cambodia, displacing over 600,000 Cambodians. France then exported *quadrillage* to Algeria after their 1954 defeat at Dien Bien Phu in Vietnam.[103] The *quadrillage* camps displaced up to two million Muslim Algerians in the first four years of the war in Algeria, a war that France lost.

France appeared to actively resist learning from these experiences. The official view in France characterized the Algerian war of independence as a *sale guerre* (dirty war) that was "disturbing but not dramatic," while believing the solution simply "a matter of police action."[104] Illustrating just how little France understood the war in Algeria, consider French President Francois Mitterrand's declaration on 5 November 1954 that in Algeria, "The only possible negotiation is war."[105] These were bold words considering the fact that France had just lost its entire colonial empire in Southeast Asia to poorly equipped rebels just a few months prior. Meanwhile, FLN leader Ferhat Abbas remarked in 1958 that the war with France was "not simply a military problem. It is essentially political, and negotiation must cover the whole question of Algeria."[106] Time was to show just how much France would suffer from a rigid adherence to tradition and a disdain for anything besides convention.

Some innovations did occur despite intense resistance from senior leadership and a generally negative attitude toward change and creativity within the French ranks. French General Charles de Gaulle considered innovations such as psychological warfare and counter-guerrilla operations in Algeria to be "an abuse of the normal activities of the armed services" while other French military experts dubbed these techniques an "infantile malady."[107] Other areas of the military did manage to produce novel ways of fighting that continue to be used by most major military forces worldwide. One such area was in the use of helicopters against insurgents. By 1959, France had enough helicopters deployed in Algeria that two battalions could be airlifted in five minutes in the *bled* (countryside), the first such use of helicopters in combat anywhere.[108] It was in this context that France developed the first wide-scale implementation of vertical envelopment in counterinsurgency operations still in use in the current time. It is important to note that the Mujahidin considered the Soviet use of vertical envelopment combined with *Spetsnaz* blocking positions to be the most effective tactic against the insurgents in the Soviet-Afghan War (1979–89).[109] Western powers were not the only forces to benefit from this innovation of vertical envelopment in counterinsurgency. When still a field grade officer in Iran's Islamic Revolutionary Guard Corps, Qasem Soleimani oversaw a successful vertical envelopment operation in 1994 against Baluchi drug smugglers in Iran's southeastern desert.[110] Vertical envelopment remains a central component of special operations forces worldwide.

Despite the repeated indications that French conventional forces and counterinsurgency methods were unsuitable for the irregular war in Algeria, France seemed to move even further in that direction. By 1952 in Indochina, the French military had built "10,000 forts, bunkers, and concrete emplacements, totaling more than 5 million tons of concrete," according to an authoritative estimate shortly after the war.[111] Two years later, France had lost the war and exported its failed methods to North Africa. French military leadership chose to fight the conventional war they remembered instead of the irregular wars they faced, a mistake for which they paid in blood, treasure, and territory. Decades later, France continued preferring suppression by direct action in irregular wars over more effective options.

France's military answer to an essentially political problem in Mali in 2013 was a strategy of "precision strike, small [Special Operations Forces] teams, conventional ground forces, and naval assets"[112] in an attempt to pummel the Tuaregs into submission without actually solving the problems that gave rise to insurgency in the first place. Unsurprisingly, France's Operation Serval led to a devolution of the insurgency into what one supporter of the French strategy described as a "diffuse regional threat that emerged in the wake of the intervention."[113] France relied on conventional tactics in Ifoghas, Mali in 2013, employing high-tech armor, skilled paratroopers, veteran infantry, and close air support from some of the most advanced aircraft in existence. Meanwhile, the Tuaregs, France's target, used nothing larger than company-sized elements employing guerrilla tactics and driving rotted pickup trucks.

An illustrative example of the mismatch of conventional forces with irregulars took place on 16 February 2013. On that day, four hours of combat resulted in a French withdrawal despite intensive close air support and the use of the most advanced military technology available. This engagement involved over 5,000 French forces supported by 1,000 Chadians mounting a frontal assault resulting in 26 killed and 62 wounded, while the Tuaregs escaped, regrouped, and, fresh from their bold stand against French forces, easily gathered new recruits.[114] Despite the arc of lessons ranging from Spain in 1809 to Mali in 2013, French methods remain influential in the American approach to irregular war, not unlike the experiences of another great colonial power, Great Britain.

The British Method

The British experience offers numerous examples of state-centered approaches to irregular wars in Africa, Asia, and the Middle East. The British approach has tended to favor unconventional warfare and indirect methods in comparison to French direct action, but this preference does not exclude the latter from the British repertoire.[115] In fact, Britain relies on its direct approach of relative superiority in conventional force employment on one hand and the indirect approach of small-scale raids and unconventional warfare in special operations force employment on the other. In the conventional context, British forces used their relative superiority at the Battle of Omdurman on 2 September 1898 during an engagement consisting of 25,000 British and indigenous African soldiers against 52,000 Mahdist irregulars. The British used technically advanced weapons and tactically superior employment of forces to produce a severely lopsided outcome of 48 British soldiers killed compared to about 12,000 Mahdists dead.[116] A few years later, the 1916 Easter Rebellion in Northern Ireland was waning until Britain executed 15 Irish nationalist leaders and implemented a counterterrorism campaign, which, combined with Irish nationalist fervor, ensured the Irish Republican Army (IRA) would be "provided with a set of holy martyrs,"[117] as one historian put it. Britain had yet to sort out how the direct and indirect approaches should best be employed against irregular forces. It was in actions like these that British forces developed their approach to irregular wars.

These developments ranged from genuine tactical innovations to less dignified methods, depending on the resistance faced abroad and the force of domestic politics at home. Britain developed the first concentration

Chapter 6. Conventional Forces in Irregular Wars   103

camps during the Boer War (1899–1902) and then employed this concept again during the Malayan Emergency (1948–60), predating the French use of this technique in their own *quadrilles* (grids) by fifty years.[118] Like France, Britain also innovated in aerial combat. The British employed close air support in one of its earliest recorded uses during the campaign to suppress the Yazidi rebellion against the British Protectorate in Iraq in 1922.[119] The Yazidis succeeded in shooting down two of the three Royal Air Force aircraft sent to carpet bomb the villages in which the rebels lived. Instead of bombing the Yazidis into submission, the resistance was emboldened and new supporters from other Yazidi tribes soon joined the rebellion. One of the Yazidis that shot down the first aircraft provided the feeling among the rebels during his oral account of the conflict: "The firing of Mausers, the noise of automatic rifles and aircraft stirred up the world.... I didn't allow that warplane to fly over in peace."[120] Despite these and other counterproductive methods, the British were at the forefront in developing ways to fight in irregular wars, especially in crafting doctrine for what is now known as unconventional warfare.

The British innovation of unconventional warfare as a distinct operation came into its own in the Middle East. Peter Kemp, a prominent member of Britain's Special Operations Executive, noted an important divide between unconventional warfare and counterinsurgency operations, with the first being a clandestine tool of states in irregular wars, and the second an overt use of conventional forces attempting to suppress insurgencies.[121] On the unconventional side, the British-supported Arab Revolt (1916–18) was designed not to oust German and Ottoman power from bases in the Middle East or to solve an irregular war problem but was instead conducted to fulfill a dream held among many leading Britons, including T.E. Lawrence, of "biffing the French out of all hope of Syria."[122] Britain quickly moved away from the indirect approach of unconventional warfare as the twentieth century wore on, favoring conventional force employment and counterinsurgency in Cyprus, Malaya, Palestine, and elsewhere.

The number of conventional British troops in these conflicts is telling. By February 1956, there were 22,000 British forces and 5,000 police against 273 EOKA troops in Cyprus, while that number increased to 43,000 British forces by 1958 to combat less than 1,000 EOKA forces,[123] the latter having only 100 automatic weapons, 600 hunting rifles, and "a handful of men."[124] Britain committed between 140,000 to 250,000 troops in Malaya against between 2,000 to 10,000 mostly Chinese irregulars.[125] Contrary to the bright picture of their performance some have painted, the British paid high prices for even small successes.[126] In 1954 for example, for every MRLA fighter killed or captured, Britain expended 1,500 man-days of patrols or ambushes, expended 1,714 artillery shells, 857 mortar rounds, and 57 aerial bombs.[127] Interestingly, British counterinsurgency strategy succeeded in Malaya but failed in Cyprus.[128] In Ireland, Britain devoted 43,000 regular military troops, 10,000 Royal Irish Constabulary (RIC) forces, 1,500 temporary "cadets" of RIC auxiliaries, and thousands of "Black and Tan" irregulars to conduct its counterterrorism campaign against the IRA in 1919. Nearly a century on, Britain's MI5 spent 44 percent of its budget to fight less than 600 IRA members in 1993.[129] While Britain and France were developing counterinsurgency and unconventional warfare concepts of operation for irregular wars in the first half of the twentieth century, the U.S. was in its own laboratory of experience in Latin America and Southeast Asia.[130]

American Force

The U.S. Marine Corps *Small Wars Manual* relegated irregular conflicts to places where foreign governments were "unstable, inadequate, or unsatisfactory" to American interests and that were awaiting an American force to "suppress lawlessness or insurrection."[131] However, most of the 180 U.S. Marine Corps interventions in 37 countries between 1800 to 1934 were conducted to protect economic and private interests rather than for any moral objective.[132] As one leading historian of American counterinsurgency experiences put it, "In American military circles the view is still widely held that guerrillas are just bandits who can be handled only with bullets."[133] The approach to irregular wars in the U.S. continues favoring a long-established strategy of suppression above any other.

The surprisingly consistent American approach to the Vietnam War is telling, given the clear indications the approach was failing. An important element of the American strategy in Vietnam from 1950 to 1964 was to form indigenous conventional units to fight insurgents because American military doctrine at the time taught that such tactics were a "lesser included capability" of conventional warfare.[134] This approach was costly, inefficient, and counterproductive. In one month of 1964 alone, 72,794 operations yielded only 406 enemy contacts, while CIA estimates identified enemy contact in under one percent of over two million combat-seeking patrols and other small unit actions from 1960 to 1968 in Vietnam.[135] The ineffective nature of American and indigenous conventional presence patrols in Vietnam is illustrated by a news item from 12 April 1964, in which 5,190 patrols were conducted in South Vietnam that week, with only 70 of those experiencing any form of enemy contact, or 1.3 percent.[136] At the same time, the American use of unobserved indirect fire toward the end of the war rapidly increased the base of support for North

Chapter 6. Conventional Forces in Irregular Wars 105

Vietnamese forces.[137] In fact, 85 percent of American or American-led indirect artillery fire was unobserved in the latter stages of the Vietnam War, not unlike the French bombardment of villages during the Druze Revolt in Syria (1925–27).[138] This was not the only parallel to French practices, however.

The U.S. also drew upon earlier French colonial methodologies. These included clear and hold, pacification, cordon and search, "hearts and minds," *tache d'huile*, *quadrillage*, search and destroy, sustained aerial bombing, and "hut hunting," which were all ineffective overall.[139] The perceptions of success felt following tactical or operational achievements on the battlefield within narrowly defined military objectives ignored their negative long-term implications. The strategic hamlet concept is an illustrative example.

The strategic hamlet program, begun in Saigon in 1962, was designed to forcibly relocate Vietnamese villagers considered by U.S. forces to be "vulnerable" to the North Vietnamese. The program had a paradoxical effect on its roughly 12,000 fortified hamlets. They quickly became known as "the stockade hamlets, with their barbed wire and concrete pill boxes," as they took on a "concentration camp character … instead of winning the confidence of the villagers, the program further alienated them."[140] Toward the end of America's war in Vietnam, North Vietnamese General Vo Nguyen Giap observed how "The Pentagon appears incapable of understanding the dynamics of a people's war."[141] This failure to understand did not end with Vietnam.

The American wars in Iraq (2003–11) and Afghanistan (2001–21) provide numerous examples of this failure to learn from the past, while themselves providing a wealth of information in military experience from which to draw. Despite this, President George W. Bush's goal to "Root out and destroy"[142] the Iraqi insurgency paralleled General William Westmoreland's vision for defeating the insurgency in South Vietnam, which Westmoreland framed simply as "Firepower."[143] The view of overwhelming fire superiority as a substitute for effective military strategy remains a potent force in the U.S. military policy community. The influential author Stephen Biddle attempted to reduce behavior in war to a matter of technology, manpower, and politics, a view epitomizing the "firepower" mentality General Westmoreland promoted in the carpet-bombing of Vietnam and General Curtis Le May emphasized in the firebombing of major urban centers in Japan during the Second World War.[144] Despite this tradition of preferring airpower as a solution to defeating insurgency, airpower was not a decisive factor in coercing adversaries, from regional conflicts such as Italy against Ethiopia and Libya to great powers like the Soviet Union in Afghanistan and the U.S. in Vietnam.[145]

The fixation on a combination of firepower and overwhelming force as a cure for all ills is not a new problem, and the U.S. has had plenty of opportunities to learn from other states. Five decades prior to the 2003 American invasion of Iraq, the Dutch were losing their own war far from home. Javanese insurgent leader Simatupang noted that the Dutch "could not understand the aspirations that constituted the driving force of our struggle ... it was difficult, of course, for these Dutch military men with their conventional military training to understand the forces they now faced."[146] Ironically, the Dutch had such a complete intelligence picture that they knew more on the Javanese order of battle than the resistance leaders knew of their own organization. And yet the Dutch still failed.

A decade later, the U.S. was embroiled in a war not unlike what the Dutch faced in Indonesia. North Vietnamese defector Le Xuan Chuyen observed that "we are fighting in our own country and know the people, the language, the terrain, and the weather of the operational area. You Americans have come halfway around the world to a country you don't know, a people you don't understand, and a terrain which is totally different than your own. These are all things you have to learn but which we know from birth. It is a natural advantage you cannot compete with."[147] This observation draws a link between the Dutch failures in Indonesia and the American failures in Vietnam: having total access to information and technology does not translate to total victory in a war against the people.

Precise articulation of mission planning factors such as the enemy order of battle, weather, terrain, time, and so on do little to explain the causes of the conflict and potential solutions to the problem underlying the people's will to resist. This disconnect also applies to the conventional force's conception of cost in terms of the right places to apply pressure, the types of operations to conduct, and the desired outcomes of these. Imposing costs is an effective strategy only to the extent that all belligerents share value judgments on the targeted quanta, e.g., oil, economics, weapons, labor, public opinion, and so on. The adversary is not bound to accept any definition of cost that the state assigns, no matter how forcefully the state may define it.

It should be clear by now that small tactical actions have strategic effects in irregular wars. A few more examples will help underscore this point. In April 2003, a month after the U.S. invaded Iraq, a U.S. Special Forces colonel learned the hard way how his misreading of the human terrain there could produce new enemies. This officer removed the Iraqi flag from the Nineveh Governorate building and replaced it with the American flag, causing residents of Mosul to mass "in fury over this flagrant insult."[148] In another episode, U.S. soldiers had conducted numerous forced entries of homes and searches of civilian women and children

Chapter 6. Conventional Forces in Irregular Wars 107

perceived by the residents to be overly invasive, leading an influential imam to deliver a fiery sermon in Tikrit in 2004, making "a call for jihad to defend the honor of men and women who had been humiliated by the behavior of American soldiers."[149] A strategic analysis of irregular warfare for American defense policy conducted in 2010 found that the U.S. defense establishment had no proven method of modeling the dynamics at play in irregular wars, and yet programmatic decisions and major force design decisions proceeded apace with no actual consensus on who the enemy was or how that enemy should be defeated.[150] One military theorist directly involved with policy formulation at that time observed that "since Americans hated the experience of military occupation in Iraq and Afghanistan—and hope never to do it again—we don't plan to engage or prepare for these types of conflict. Instead, we will structure our forces for the wars we want to fight."[151]

The misunderstanding and mischaracterizations of the U.S. wars in Iraq noted above are supported by a study following the 2003 invasion. That study found "The prolonged occupation of Iraq, and the failure to reconstitute a functioning government able to garner widespread legitimacy and police its borders, generated motivations for, and enhanced the ability of, terrorist groups to form and fight."[152] Marine Lieutenant General George Smith urged that military leaders should promote "a willingness to fail, learn, and adapt quickly during periods of relative peace" to remain competitive and successful in future conflicts, yet even with these words, lessons are forgotten far faster than they are learned and reflected upon. However, the failure of French, British, American, Israeli, and other conventional forces in irregular wars is not simply a matter of poor operational design and insufficient planning.

States fundamentally misunderstand or ignore the causes of irregular wars, leading to strategies that confuse means with ends along with virtually unbounded military commitments to those conflicts. A chief transgression states commit in these conflicts is entangling terrorism with irregular warfare. Most contemporary military scholars frame terrorism as "the weapon of the weak against the strong, [which] works by blurring boundaries between combatants and non-combatants."[153] Those using this framework cannot untangle "terrorism" on one hand and "irregular warfare" on the other, despite the former being a way to describe a tactic that anyone could use and the latter being the conduct of an irregular war.[154] More usefully, terrorism and sabotage can be considered subversive activities, to differentiate this cluster of tactical actions from the operations insurgents conduct.[155] It must be remembered that terrorism, espionage, subversion, and sabotage are simply tactics conducted against particular targets and have nothing to do with the party responsible for their execution. Put another way, these four activities are dependent upon their targets, not the executors. In this way, terrorism targets civilian pathos, espionage targets protected information, subversion targets organizations, and sabotage targets physical installations. States commit these four acts all the time, far more often than insurgents, and often against other states. Terrorism, espionage, sabotage, and subversion have as much to do with causes of irregular wars as the U.S. firebombing of Tokyo did with the cause of the Second World War. Tactics are not causes.

The failure to understand the causes of irregular wars is a costly mistake. The Iraq Study Group Report found that "Congress has appropriated almost 2 billion [dollars in 2006 alone] for countermeasures to protect our troops in Iraq from [improvised explosive devices], but the administration has not put forward a request to invest comparable resources in trying to understand the people who fabricate, plant, and explode those devices."[156] This failure to link cause, effect, and appropriate methods has been costly in both dollars and lives. U.S. spending on the Global War on Terror during the period of 2001 to 2019 cost American taxpayers 4.933 trillion dollars, without counting future obligations, commitments, or program costs.[157] Just one year later, the cost had increased to 6.4 trillion dollars.[158] Expenditures just in Afghanistan alone cost over 2.2 trillion dollars—more than 300 million dollars a day, every day, for 20 years.[159] The Global War on Terror displaced 37 to 59 million people from only the top 8 of the 24 countries most affected by the resulting American-led combat operations.[160] The number of displaced persons "exceeds those displaced by every war since 1900, except World War II … deaths and injuries number in the millions."[161] Additionally, 800,000 combatants have died, and 4 to 12 million civilians have been killed as a result of the U.S.-led campaign to kill "terrorists," leading one study to conclude that "the legitimacy and efficacy of war should be questioned more than ever given nearly two decades of disastrous outcomes."[162] Despite these expensive lessons, the U.S. was still directly involved in counterterrorism activities in 80 countries in calendar year 2018 alone.[163] Half of those 80 states were hosting U.S. forces for counterterrorism operations, while 65 were undergoing counterterrorism advise and assist programs, 26 were hosting U.S. counterterrorism exercises, 14 had troops actively engaged in or directly supporting combat operations in those countries, and 7 were active theaters of U.S. airstrikes from manned and unmanned platforms.[164] The lesson is this: countering insurgencies is an operation, not an objective, and suppression often strengthens resistance. American defense policy has long placed primacy on suppression operations such as counterterrorism, counterinsurgency, and direct action.

History provides numerous examples of the suppression strategy in irregular wars leading to long-term failure notwithstanding perceived short-term benefits to the state. The example of Soviet operations against Ukrainian resistance in 1918 illustrates how a combination of conventional military forces, intentional spreading of diseases, deportation, targeted killing of ideological leaders, forced migration, and imposed famine achieved short-term objectives while leading to costly consequences over time.[165] Russian attempts to subdue Chechen resistance by assassinating Chechen resistance leader Dzhokhar Dudayev in 1996 had a negligible effect on that resistance movement. Similarly, Israel's assassinations of Hamas and Hezbollah leaders since the Israeli invasion of Lebanon in 1982 illustrate a more precarious reality: attempts to destroy a resistance movement with a decapitation strategy often paradoxically strengthen the target movement significantly. In one of many recorded speeches, al-Qaeda leader Osama Bin Laden discussed his motivations for conducting attacks against the West, calling attention to Israel's invasion of Lebanon in 1982 and its effect on his later motivations: "The situation was like a crusade ... to punish the oppressors."[166] But one need not look to the extreme measures the Russian forces or Israeli assassins took for lessons to learn— Western examples of the suppression and direct action approaches to irregular wars are perhaps more instructive.

The failure of numerous American and British attempts to transform Libya by assassinating Muammar Qaddafi beginning in 1986 proved correct political scientist John Mearsheimer's observation that "decapitation is a fanciful strategy."[167] It is unsurprising that the assassinations of Anwar al-Awlaki, Osama bin Laden, Abu Bakr al-Baghdadi, Abu Musab al-Zarqawi, Imad Mughniyeh, Ayman al-Zawahiri and numerous other targets in the Global War on Terror achieved more in satiating a thirst for vengeance in Western living rooms than they did in eliminating the insurgencies these individuals represented. Public opinion, especially in liberal democracies, can be a major obstacle to selecting appropriate methods for states to approach irregular wars since attempts to ameliorate constituents often help exacerbate the conflicts.

The Baghdad Pact of 1955 reflected President Eisenhower's desire to show the American people that he was committed to fighting Communism. The Baghdad Pact featured a counter-subversion committee directly addressing conflicts in Iran, Iraq, Turkey, and Pakistan, all framed as Soviet puppet movements rather than genuine agitations for independence or representation.[168] President Eisenhower believed that "Russia's interest in the Middle East was solely that of power politics" according to his introductory statement to the Baghdad Pact in 1955. The U.S. responded with generous funding to those regimes opposing Russian support through the 1957 Eisenhower Doctrine, without concern for their right-wing authoritarian and anti-democratic governments. No attempt was made in the Baghdad Pact to assess the reasons civilian populations in the Third World were agitating against authoritarian or colonial regimes—this agitation was simply accepted as a function of global Communism, even if this was not the case.

President Johnson followed this trend of choosing strategies that were domestically popular but were actually known to make the conflicts worse. In the 1960s, the Johnson administration employed a conventional strategy of overwhelming firepower known in advance to be ineffective against the insurgency in South Vietnam. Despite this foreknowledge, Operation Rolling Thunder was approved because of high domestic support for the bombing campaign from the voting public in America.[169] A few years before that bombing campaign commenced, President Kennedy's foreign policy in the Third World consisted of "Reform first but tough repressive counterinsurgent measures second."[170] One contemporary observer assessed that Americans sought "a full-scale crusade against Communism in the Orient."[171]

Closer to home, the U.S. established the School of the Americas to manage the counter-revolutionary responses for El Salvador, Guatemala, and Nicaragua. The U.S. was not alone in investing heavily in kinetic options for suppressing insurgency in Latin America with, for example, Peru spending 10 million dollars to fight less than 100 insurgents in 1965.[172] During President Reagan's administration, the School of the Americas cost U.S. taxpayers 600 million dollars on average per year from 1980 to 1990.[173]

Suppression and Destruction

The examples of France, Britain, and the U.S. illustrate the inadequacy of suppression and destruction strategies in irregular wars, as well as the high costs for these failures in lives, money, and prestige. France favored direct action and targeting individuals in its approach to countering insurgencies over the previous century, while Britain relied on numerical superiority and unconventional warfare. The U.S. tended to favor French methods over British, while eventually incorporating some of the latter but still retaining the preference for suppression over any other strategy. German theories also permeated U.S. military thought especially for irregular warfare engagements and chasing the dream of the "combat hunter." Director of Central Intelligence William Colby commented in the 1970s that the role of a state in an irregular war was to "Put pressure on an enemy or to strengthen an insurgent,"[174] illustrating the view conventional forces often take to irregular wars: insurgents may be either suppressed or destroyed, and there are no other options.

This preference is not limited to the great powers. During Israel's failed 2006 invasion of Lebanon, Israeli forces outnumbered Hezbollah by at least 15 to 1, with some estimates placing the imbalance at 65 to 1 depending on the method of calculation.[175] The outcome of the battle was clear: Israel could not defeat the tiny Hezbollah formations even though Israel possessed the most advanced weaponry available in the world, enjoyed full support from the U.S. as a third party, and fielded excessive numerical superiority among a range of other advantages. The way states and other actors intervene in irregular wars indicates that suppression and destruction may not be an effective course of action in such conflicts.

Not all states remain attached to suppression in irregular wars. India's counterinsurgency doctrine, consisting of five components, was employed in Assam and Kashmir with mixed results.[176] This doctrine limits force, isolates insurgents, dominates a target area, garrisons troops in contested areas, and seeks political rather than military solutions to the conflict.[177] This counterinsurgency approach is articulated in the Indian Armed Forces *Doctrine for Sub-Conventional Operations*, which defines sub-conventional war as "all types of armed conflicts above the level of peaceful coexistence and below the threshold of war, such as militancy, insurgency proxy war, and terrorism apart from border skirmishes."[178] While the Indian military still has some way to go, it has attempted to take its own path in handling its internal defense issues in Assam and Kashmir rather than adopting the Western approach wholesale.

Third-Party Intervention

Foreign intervention is common in irregular wars for several reasons, not least of which is to prevent state death. Indeed, the birth of states from irregular wars is perceived to be an existential threat to the international order because this process upsets the delicate balance existing states struggle to manage. The balance is sensitive to violence against states. The types of violence most threatening to states can be placed along a scale sometimes called a "scattered attack" continuum.[179] This continuum consists of uncontested shows of force, skirmishes among equally balanced forces, and unbalanced forces contesting through resistance. The continuum is depicted in Figure 5.

The continuum of force balance and force display is fundamentally about the degree of freedom potentially exercised by the forces of a state, usually their military or security forces, without resistance to

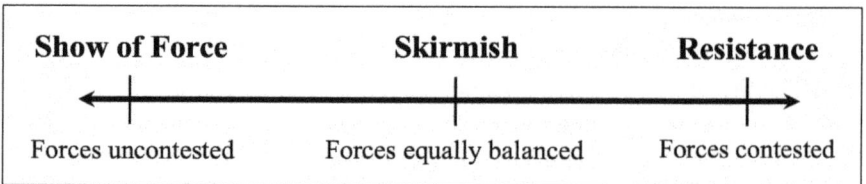

Figure 5. Continuum of Force Balance and Force Display.

their activities. For example, a military parade in a capital city is often an uncontested show of force because it does not explicitly seek a rival parade from another state's military or security forces. Meanwhile, a show of force that is equally balanced with an opposing military or security force is frequently seen in the dead space between the Forward Line of Own Troops (FLOT) and the Forward Line of Enemy Troops (FLET).[180] In cases of equal force balance and display, there is an agreement, often unspoken, that the interaction is for display purposes rather than a resistance-inducing action. Deterrence is a form of force display, express or implied, in which the objective of the display is to reduce potential resistance. As resistance to the display increases, skirmishes occur and propel the opposing forces toward a contested interaction. The ultimate state of contested forces through resistance is warfare. When this contest is between a state and an insurgency, the forces are engaged in irregular warfare.

Members of the Westphalian state system favor barriers to entry into their club of sovereign states, and great powers have the most to lose. Power, uncertainty, and fear govern state-level decisions to embark on third-party interventions in irregular wars.[181] The defense of state interests underpins third-party involvement in irregular wars, and the three primary approaches affected states take are intervention, accommodation, or abstention during an insurgency or revolution, with intervention appearing most frequently.[182] States intervene to protect their own interests and those of other states rather than for any claimed moral purpose, regardless of how loudly states defend those morals in public fora.[183] More precisely, the inclination for states to defend perceived interests explains why those states may erect barriers to state formation while attempting to generate predictable balances of power by supporting or preventing *coups d'état*, stability operations, counterterrorism, counterinsurgency, foreign internal defense, unconventional warfare, and other forms of third-party intervention in the affairs of a *de jure* state.

The Russian Civil War (1916–26) is an important example of third-party intervention in a modern irregular war during a moment of liminal sovereignty resulting from sovereign dysfunction. Third-party support took several forms in this case, including unconventional warfare,

Chapter 6. Conventional Forces in Irregular Wars 113

resulting in what the commander of the French component later admitted was "the complete failure of a ridiculous adventure."[184] The Allied involvement with the anti–Bolsheviks, White Russians, and others was bolstered through the creation of the Volunteer Army, all supported more to balance against the Axis Powers than to stymie a Soviet rise. However, the publicly stated objective for the Allied support was to "Achieve isolation of Russian Bolshevism with a view to bringing about its destruction," according to French Prime Minister Georges Clemenceau.

Regardless of the objective, the Allies provided unprecedented advice, assistance, and materiel to the forces resisting the Bolsheviks. Allied military support included an expeditionary force of 6,300 British, 14,000 American, 1,800 French, and 4,000 Canadian forces landing in various areas around Russia including Siberia, the Caucasus, and the Baltic states.[185] Financial support included £50 million, inflation adjusted to about £3 billion in 2022, while Britain provided air and artillery support during combat operations. The British also conducted direct action raids on Soviet naval posts while providing additional military aid and assistance to Latvia, Lithuania, and Estonia. The British, French, and American effort included two separate unconventional warfare components, one for the White Russians and the other for the Volunteer Army, depicted in Table 1.[186]

Table 1. Allied Materiel Supplied in Russian Civil War (1916–26)	
To White Russians	*To Russian Volunteer Army*
500,000,000 cartridges	500,000,000 rounds for small arms
200,000 uniforms	500,000 uniforms
2,000 machine guns	6,100 machine guns
400 heavy guns	1,200 heavy guns
155 aircraft	100 aircraft
Tanks (unspecified quantity)	74 tanks
	2,000,000 artillery shells
	629 trucks and ambulances
	270 motorcycles
	6 armored cars
	12 general hospitals, 500 beds each
	25 field hospitals
	Radios, electronics, and engineering equipment

The interested third party is not always successful and intervention often exacerbates the very causes that led to the irregular war in the first place. This is especially true if military support overshadows the needs of civil society.[187] Third party intervention, no matter how massive, committed, or intense, cannot take the place of popular support for the side the third party has intervened to prop up.[188] This fact was not lost on French revolutionary leader Maximilian Robespierre, as he opined in 1792 that

> The most extravagant idea that can take root in the head of a politician is to believe that it is enough for one people to invade a foreign people to make it adopt its laws and constitution. No one likes armed missionaries; and the first advice given by nature and prudence is to repel them as enemies.[189]

Despite France having gained such insights in their own revolution, the French experience in the Russian Civil War was "a complete fiasco," according to one French general involved in the fighting.[190] The French were "Poorly informed about the chaotic political conditions in Ukraine," which led to French forces "facing Red Army units, Ukrainian separatists, and several hostile partisan groups. Efforts to convince the [White Russians] and the Ukrainians to join forces against the Bolsheviks proved fruitless, and the expeditionary force was forced to withdraw in April [1920]."[191] The powerful force clouding the judgment of France and indeed many other states was the desire for escalation control. Theaters of irregular war incentivize third parties to intervene below direct confrontation with other states, as this indirect, clandestine, or covert confrontation exerts escalation control for the same states in other theaters.

The Allied intervention in the Russian Civil War is one example of this behavior, while Nazi and Soviet intervention in the Spanish Civil War (1936–39) is another.[192] Despite the irregular war context of the Spanish Civil War, the conduct of the war itself, its warfare, was generally conventional. This was why Peter Kemp, a British national participating as a foreign fighter with the Nationalists, received the recommendation from a peer to use the *British Field Service Regulations* in his military actions in Spain, since the war in 1936 was fought "more or less according to textbook principles."[193] In another example of conventional warfare in an irregular war, Yugoslavian irregulars successfully confronted nearly 12 Axis divisions from January to June 1943 in Case White and Case Black, prompting Winston Churchill to observe that "the guerrilla forces in Yugoslavia and Albania are containing as many German divisions as the British and American armies put together."[194]

States are often enticed to use irregular forces to conduct wars on their behalf to maintain a degree of separation from the conflict. This behavior of using non-state proxies to avoid state-on-state confrontation is known as the "limited war" theory of state secrecy.[195] Limited war theory contains a useful model for framing third-party intervention from a major power. This model considers other external powers, the internal/primary factions, the territory upon which the conflict occurs, and the periphery that can also facilitate any of the participants in an advantageous manner.[196] Considering the use of non-state proxies as a method to avoid confrontation between states helps to shed light on the different motives for overt and covert third-party intervention in irregular war. This also demonstrates that escalation from third-party intervention can be vertical or horizontal, with the first arising from forms of combat power and the second from territorial expansion of the conflict area.[197] A third form of escalation can emerge through a phenomenon called "object shift," which occurs when a relatively localized, minor issue of contention rapidly escalates upward and outward during broken negotiations to involve larger, broader poles of opposition.[198] The Arab Spring and broadening of the Vietnam War are two examples, both containing the important act of self-immolation, a display certain to provoke object shift.

It follows that weak, threatened, or failing states can use the threat of escalation to persuade powerful patrons to intervene either directly or with surrogate forces. The incentive to limit the appearance of war while supporting its conduct has been a common occurrence since the earliest days of the Westphalian system, though this activity predates that system. Venice provided lethal aid to the Druze resistance in Syria during the first Ottoman conquest (1516–21) and again during the second attempt (1523–86). In fact, the classical Arabic word for a musket or rifle is *bunduqiyya*, the Arabic name for Venice, the chief supplier of these weapons to the region at the time.[199]

Sometimes states enable irregulars through unconventional warfare simply to make the conventional aspect of the war more difficult for the opposing state or to appear overtly neutral while participating covertly. To make things difficult, the activities of pro–Bourbon guerrillas in Spain in 1707 made the allied presence in Madrid untenable during the War of the Spanish Succession (1701–14).[200] To remain overtly uninvolved during the War of Austrian Succession (1740–48), France and Britain used contingents of auxiliary troops to satisfy treaty obligations for Austria and Bavaria while remaining officially neutral toward each other and to avoid formal declarations of war.[201] After ascending to the French throne in 1774, King Louis XVI instructed his ministers "to meddle adroitly in the affairs of the British colonies; to give the insurgent colonists the means of obtaining support of war, while maintaining the strictest neutrality."[202] However, states have a long tradition of miscalculating the degree of support their surrogates offer. The British mistook the relative indifference in the

American South during the American Revolution as active support to the British war effort against the insurgency in the colonies. Consequently, the British miscalculation led to extensive delays in establishing a base in the friendly city of Charleston, with the project dragging on for almost two years.[203] Foreign irregular forces and states have also long relied on third countries for their confrontations, coopting ongoing conflicts to resolve their own violent disputes, as occurred in Spain.

Third-party involvement in the Spanish Civil War (1936–39) began immediately as private citizens of Nazi Germany, Fascist Italy, neutral Portugal, and many other independent foreign fighters flocked to the Nationalist side. On the Republican side, the Soviet Union, France, and Mexico lent their own forces, while independent foreign fighters quickly arrived. In some cases, foreign fighters from the same countries, such as Russia and Great Britain, rushed to Spain to fight their countrymen on opposing sides. British soldier Peter Kemp and British-Soviet double agent Kim Philby joined the Republican side against the International Brigades, in which many British, Canadian, and Americans fought, including author George Orwell. The White Russians also continued warring unofficially against the Soviets after failing to regain power in Russia through their own third-party intervention in Spain.[204] Despite these interventions, the Spanish Civil War was essentially a war among contested sovereigns to dominate state capacity in Spain.

Clusters of fragile governments are a leading cause of conflicts usually requiring external support to concentrate capacity at the level of the central government to reassert state sovereignty.[205] However, no level of third-party support can overcome sovereign dysfunction without a requisite degree of coherence within the state.[206] It is for this reason that although outside support was an important factor contributing to the Viet Minh's military endurance against France, Japan, the U.S., and later, China, this was hardly necessary for their overall appeal to the Vietnamese people, especially in the north and the delta region. Indeed, South Vietnamese received far greater external support and they still lost.[207] The Viet Minh were also helped indirectly by their enemy, as the U.S. provided a significant number of weapons to the South Vietnamese that were subsequently lost to the communists. Between 1960–63 alone, the Viet Minh made a net gain of 128,682 weapons captured from South Vietnamese forces, which was about enough weapons taken to equip each insurgent in South Vietnam with one rifle.[208] Lebanon provides similar examples of third-party intervention having unintended negative consequences.

Operation Blue Bat began on 15 July 1958 when over 14,000 American forces landed in Lebanon, encircling Beirut to enforce political outcomes favorable to the U.S. and Britain.[209] Sunni and Druze groups

Chapter 6. Conventional Forces in Irregular Wars 117

participated in the 1958 uprising in Lebanon because they felt excluded from government posts and the associated influence those positions offered to the incumbents, which they considered to be a "Maronite business oligarchy."[210] Although numerous consequences flowed from Operation Blue Bat, the conditions it exacerbated were partially responsible for the Lebanese Civil War (1975–90), the Israeli invasions in 1978 in which over 2,000 Lebanese civilians were killed in Israeli air raids, and in 1982 with the major Israel Defense Force ground invasion that remained until 2000.[211] Syrian attempts to absorb Lebanon into a "Greater Syria" continued stoking divisions as well, while Cold War powers played Israel and Syria against one another in the theater of Lebanon's cities, towns, and villages.[212] A major consequence of these invasions and occupations occurred after the Iranian Revolution in 1979. In 1983, Iran's Office of Liberation Movements, renamed the Islamic Revolutionary Guard Corps-Quds Force in 1988, sent a mission to Lebanon to create a militia now known as Hezbollah out of a group of disaffected Amal Movement members.[213] The group that would become Hezbollah had a simple goal. Their objective was to violently resist the third party intervention of Israel, the American-led multinational force in Beirut, and the Israeli surrogate militias comprised mostly of Maronite nationalists fighting primarily in the South Lebanon Army within their so-called Free Lebanon State.[214]

Iran's Office of Liberation Movements was already focused on resistance to Soviet forces in Afghanistan and then turned toward Lebanon by 1981. This organization and its successor, the IRGC-Quds Force, used four components of its resistance project in Lebanon in the 1980s: proselytizing; military and ideological training of cadre, of which 60 per cent were military; funding; and recruiting resistance members.[215] It was during and immediately after this time that IRGC-Quds Force began its working partnership with Hezbollah, Hamas, Palestinian groups in Syria, and the Badr Organization in Iraq, among others.[216]

The lessons from Lebanon were forgotten almost as quickly as the operations took shape. Reflecting on the American military intervention in Haiti three decades later, General Wesley Clark revealed that the collective memory in the U.S. military had already deleted Operation Blue Bat in Lebanon: "We had learned from the experiences in Haiti—when we deployed 20,000 U.S. troops to oust the Haitian coup leaders and their infamous military-police force—that we should never look to our military to do police work ... searching for weapons and trying to enforce disarmament was a matter for police."[217] Rather than retaining difficult lessons from Lebanon, and indeed from Haiti, the U.S. went on to repeat intervention errors elsewhere, including Libya, Iraq, Syria, Yemen, Somalia, Afghanistan, and numerous other locales.

The U.S. has had ample opportunity to learn from the mistakes of third-party interventions in irregular wars, both from its own past endeavors and from those of other states. David Johnson led a RAND fact-finding team to Israel following the Israel-Gaza War of 2008 to see what lessons could be learned for the U.S. During that conflict, Israel conducted 5,650 aircraft sorties using advanced fixed-wing multi-role fighter aircraft, helicopters, and unmanned aerial systems across the densely populated Gaza Strip, killing 1,400 Palestinian civilians, and destroying over 4,000 of their homes at a cost of only 6 Israeli casualties despite deploying some 30,000 Israel Defense Force ground troops during the 22-day campaign. In contrast, the most advanced weapons the Palestinians employed against Israel were various Soviet-era small arms, rockets, and mortars, which were munitions first developed in the fourteenth century.[218] The Israeli invasion further exacerbated the conflict, damaged Israel's prestige domestically and internationally, and strengthened support for Hamas in Gaza, resulting in Hamas taking over the political institutions there. Like Israel's invasion of Lebanon in July 2006, the bombardment of Gaza was clearly disproportionate and was predicated on a questionable interpretation of military necessity.[219]

The modern usage of military necessity is based on the German idea of *Kriegsraison*, which holds that "a ruthless war is quicker, and is therefore more humane."[220] While British and American military doctrine call for the minimum force necessary, this is in practice a privilege reserved for adversaries that are states.[221] Those that are not states instead receive what has been described as "exemplary force," which is designed for "awing the population with force" rather than prioritizing redress of grievances, thus strengthening the opposition's appeal at the expense of the exemplar.[222]

Given this interpretation of military necessity, it is perhaps unsurprising that the RAND team returned from their fact-finding mission in Israel to suggest the U.S. should adopt Israel's disproportionate, expensive, and ineffective methods. They concluded that "high-end conventional ground combat capabilities are required to prevail against sophisticated irregular adversaries."[223] It was also not unexpected, albeit disappointing, that then–Defense Secretary Robert Gates promoted the use of tank maneuvers specifically developed to repel a Soviet invasion through the Fulda Gap in Europe to defeat insurgents in Iraq who had no tanks.[224] Similar decisions were made in 2017, as U.S. military advisors and Philippine security forces laid siege to Marawi, "during five months of brutal fighting involving the bombing of entire neighborhoods and thousands of civilian deaths."[225] This resulted in the forced displacement of about 200,000 residents of Marawi, which was nearly the entire surviving population. The Philippine and U.S. governments considered the operation a

Chapter 6. Conventional Forces in Irregular Wars

success since a handful of notorious Da'esh-affiliated fighters were killed. However, 95 percent of all buildings in the four-square kilometer downtown area were destroyed mostly by U.S.-Philippine airstrikes.[226] Combat operations are not the only area third-party interventions seem to produce long-term negative consequences.

Foreign internal defense, security forces assistance, and stability operations may also have negative consequences over time in the host nation if the security pillar is strengthened disproportionately to the political-economic pillars in a weak state with a dysfunctional sovereign. The training the U.S. has provided to foreign militaries and security forces has correlated to a significantly higher rate of *coups d'état* occurring following those programs.[227] There is a statistically significant increase in the likelihood of *coups d'état* in those states as the level of American military education and training increases, because those foreign military members who receive a form of education that prizes autonomy, decentralized command and control, and initiative may use those new skills against the dysfunctional sovereigns in their home states. This is compounded by the fact that some of those military members inevitably grow disaffected, less invested, and even hostile to the survival of their dysfunctional sovereign once alternative methods, means, and opportunities arise for seizing the initiative for change in under-institutionalized states.[228] In a cultural context, security forces assistance programs may impose alien American conventional forces traditions upon foreign militaries who have incommensurable gaps between their own organizational culture and that of the U.S. military.[229]

One prominent example of this process is the rise of Abdel Fatah El-Sisi in Egypt from a lowly Egyptian military officer to the president of Egypt after leading a military junta to oust democratically elected president Mohammad Morsi in 2013. El-Sisi attended the United Kingdom's Joint Services Command and Staff College in 1992 and then the U.S. Army War College in 2006. While at the Army War College, El-Sisi wrote a thesis entitled "Democracy in the Middle East," in which he frowned upon the Egyptian and American governments and described the free press as "an obstacle" to government.[230] Seven years later, he had helped to overthrow that government he had derided in his thesis, and he was now the president.

The U.S. is not alone in failing to learn from past forays into third-party interventions in irregular wars. Russia ignored its own lessons learned from the Soviet-Afghan War (1979–89), which was a contributing factor for Russia's early failures in Chechnya during the 1990s.[231] Indeed, this resistance to an institutional knowledge of counterinsurgency appears to be a common theme for conventional forces.[232] One area that the major powers have remained consistently invested, however, is unconventional warfare.

Unconventional warfare is a form of third-party intervention in which a state supports a resistance movement or militia with the goal of overthrowing, disrupting, or displacing an incumbent state. Such surrogate forces and insurgencies are actually military instruments wielded to achieve political outcomes, much as conventional warfare is wielded between states.[233] Unconventional warfare is one method nuclear powers may use to engage in conflict with other nuclear powers or their allies while keeping the likelihood of escalation to open combat to a minimum. The nuclear policies of the Great Powers during the Cold War led to "client military forces, secret services, the stimulation and supply of insurgencies, and the support of transnational terrorists"[234] all forming tunnels beneath the strong walls of deterrence, mutually assured destruction, and massive retaliation.[235] The U.S. employed such a strategy of tunneling beneath deterrence against the Soviet Union in Afghanistan in the 1980s.

The CIA would go to extreme lengths to degrade the Soviet military adventure in Afghanistan. To achieve this, the CIA began Operation Cyclone to support surrogate forces in Afghanistan against the Soviet-backed Afghan government from 1979 to 1989. Operation Cyclone was "the longest and largest covert operation" in American history, and its most expensive.[236] The lack of accountability for the CIA's surrogate forces in Afghanistan promoted the proliferation of "armed groups that have no legal standing in the country where they operate," allowing the CIA "to run operations shielded from transparency and public accountability," sabotaging the overall American and international political goals for those countries.[237] This had not changed a decade later, as that same lack of accountability in unconventional warfare exacerbated the security issues in Afghanistan and Pakistan after the American-led invasion in 2001.[238] Meanwhile, this pattern hindered America's own follow-on mission of stabilizing the new government with rule of law, legitimizing security institutions, and creating mechanisms to form institutions capable of representing the Afghan public. However, accountability alone does not always lead to a moral outcome in unconventional warfare operations, as it may instead be conducted for reasons of *realpolitik* to the benefit of a neighboring state.[239]

The Russian invasion of Ukraine in 2014 is an illustrative example. Russian unconventional warfare activity in Ukraine in 2014 was not focused on overthrowing the government in Kiev but instead undermining the legitimacy of Ukraine's political institutions and thereby enabling Russia's annexation of Crimea followed by Russia's less successful actions in the Donbas region.[240] Russia used *spetzpropaganda* (special propaganda/

disinformation) and *maskirovka* (military deception) as weapons to create favorable conditions for executing foreign policies other states opposed.[241] From this experience, the Russian Armed Forces Main Operational Directorate under Andrei Korybko developed a new theory of warfare in 2015, which gave two avenues for regime change: "The non-violent or 'soft' overthrow via color revolutions and the violent overthrow of unconventional warfare."[242] Russia feared the West would use the color revolutions against states within the Russian sphere of influence, and the Russian solution to this problem was widespread unconventional warfare in both the physical and information spaces.

The portion of Korybko's theory pertaining to unconventional warfare reflected the Russian experience in Ukraine in 2014. Russia learned four lessons from its unconventional warfare operation in that conflict.[243] First, operations should be deniable. Second, direct action should occur through surrogates. Third, information is a weapon that should supersede actual violence in its effects. Fourth, preparation of the environment is critical, including preparation of the information environment. The Russian annexation of Crimea in 2014 and intervention in Georgia in 2008 were exceptions to the norm, as irregular wars are not always a matter of states intervening in the affairs of other states for the simple purpose of rational self-interest. Despite this, Russia again repeated the formula in Syria to stabilize its political and economic interests there following the rise of Da'esh and the subsequent growth of U.S.-backed militias in the region to counter that threat.

Militias are an important spoiling factor in what may otherwise be a straightforward third-party intervention. Militias may exist in parallel or opposition to an insurgency. This may also occur in irregular theaters of conventional wars, such as in the Second World War with the French *Maquis* being challenged by another irregular force, the *Milice française*, which was backed by Nazi Germany.[244] One useful definition of militias is "armed groups that operate alongside regular security forces or work independently of the state to shield the local population from insurgents."[245] It is important to differentiate here between "state-manipulated" and "state-parallel" armed groups, as the latter are horizontally associated with states, acting as parastatal organizations rather than vertically subordinate to states.[246]

Applying this distinction in the context of the Lebanese Civil War, the South Lebanon Army was state-manipulated while Hezbollah is state-parallel. Militias in Lebanon are hardly novel, however. Other examples include France's recruitment for *Les Troupes spéciales du Levant* (Levantine Special Troops) from indigenous ethnic and religious groups in Syria and Lebanon from 1916 to 1946, especially targeting Armenian Christians for its ranks. France sought to use *Les Troupes speciales* as the kernel of a new military force for post-war Syria and Lebanon secretly arranged with the British in the 1916 Sykes-Picot Agreement.[247]

In Africa, numerous militias operate in various roles across the continent. In Sudan in 1994, Khartoum's military used the state-manipulated Lord's Resistance Army militants based in Uganda to harass SPLA base areas from the south, while conducting air strikes on hospitals and refugee camps housing SPLA family members. Meanwhile, Khartoum also used state-parallel *mujahidin* irregulars in the north to fight SPLA-aligned northern armed groups who opposed Khartoum for their own reasons.[248]

Third parties themselves need not be combatant groups in a military sense to intervene in irregular wars. Corporations have backed foreign military interventions since the earliest days of free market capitalism, starting in 1551 in Britain with the Russia Company, the first business incorporation and joint stock company in history.[249] Multinational corporations are important non-state actors, while other transnational organizations have instigated third-party interventions in irregular wars, such as the Organization of the Petroleum Exporting Countries, the Catholic church, and familial confederations such as the al-Shammari tribe in Syria and Iraq. This link between the rise of corporations and foreign interventions is well-documented.[250] Some of these entities have held more capacity and exercised more sovereignty than the states in which the influence of these groups was felt, such as the Dutch East India Company, "both merchant and king,"[251] and the Newmont Mining Corporation in Indonesia.[252]

Mining companies in particular have had significant influence as both non-combatant and non-state actors in irregular wars. The De Beers Diamond Protection Force went to Sierra Leone in 1936 to violently protect British mining interests under the British Sierra Leone Selection Trust, eventually growing into a massive private militia that dwarfed the actual colonial government.[253] Similarly, French irregulars under the notorious French officer Roger Trinquier defended Katanga's secession from Congo (1960–62) at the behest of the *Union Minière du Haut-Katanga* (Mining Union of Upper-Katanga).[254] However, the United Nations Security Council agreed there would be no independent Katanga in central Africa.[255] The United Nations exerted sovereignty over Katanga through its first "peace enforcement" operation, named Operation Morthor, which included displaying corpses in public to dissuade further resistance to rule from the United Nations–backed Congolese central government.[256]

Like mining companies in Africa, fruit and coffee companies in Latin America spurred third-party interventions in irregular wars there. The cycle of "foreign interventions and local resistance" marring Central America especially from 1899 onward began with United Fruit

Company.[257] Meanwhile, the "coffee oligarchy" and associated *caficultura* (coffee culture) emerged in Costa Rica and elsewhere in Central America in the latter half of nineteenth century.[258] The Roosevelt Corollary of 1903 expanded the 1823 Monroe Doctrine, paving the way for American hegemony in Central America to support these activities, with special focus toward concentrating coercive power for corporate interests, including military intervention and the idea that "it was almost always 'good for business to collaborate with dictators' in Central America."[259]

The third-party interventions and support granted to prevent state death for oppressive governments in Asia, Africa, the Middle East, and Latin America largely contributed to the growth of violence in irregular wars rather than preventing them. In his analysis of Operation Odyssey Dawn in Libya in 2011, Christopher Chivvis held that "intervention will remain one of the most difficult and vexing international security issues U.S. and allied policymakers face."[260] As the foregoing examples show, history has done much to provide ample lessons to ensure this international security issue could be less difficult and less vexing. However, as long as states continue ignoring history, they will be doomed to an ongoing cycle of complex insurgencies, costly state interventions, and lengthy irregular wars.

Part III

A Theory of Irregular War

Chapter 7

Irregular War Conditions

> To attempt to arrest a revolutionary movement by means of deployed armies is like trying to use a broom to sweep back the tide. The only way to take action is to eliminate the causes.—WOODROW WILSON[1]

Conditions of Social Order and Sovereign Dysfunction

The interactions between a sovereign power and civil society may lead to irregular war if the sovereign becomes sufficiently dysfunctional in its relationship to the social order. This interaction is not simply the story of coups or juntas, authoritarians or despots, though irregular war conditions may occur in these situations. Likewise, a baronial rebellion such as the signing of the Magna Carta at Runnymede in 1215 is different than a violent transformation of the social order because the former seeks redress through negotiation against the autocrat rather than against the system of government itself.[2] Instead, the instigating condition operates in the interaction between a dysfunctional sovereign and the primary elements of the social order, something Leon Trotsky dubbed "the fundamental premise of a revolution."[3] This is a condition in which "the existing social structure has become incapable of solving the urgent problems of development of the nation."[4] Sovereign dysfunction is only a variable leading to conditions in the social order favorable to irregular war and does not always guarantee such a conflict will occur. However, a fundamental transformation of the social order in an irregular war is unlikely without it.

John Locke captured this important point when describing social revolution, which he considered the remedy to great mistakes of government:

> I answer such *Revolutions happen* not upon every little mismanagement in publick affairs. *Great mistakes* in the ruling part, many wrong and inconvenient Laws, and all the *slips* of humane frailty will be *born by the People*, without mutiny or murmur. But if a long train of Abuses, Prevarications, and

Artifices, all tending the same way, make the design visible to the People, and they cannot but feel, what they lie under, and see, whither they are going; 'tis not to be wonder'd, that they should then rouze themselves, and endeavour to put the rule into such hands, which may secure to them the ends for which Government was at first erected; and without which, ancient Names, and specious Forms, are so far from being better, that they are much worse, than the state of Nature, or pure Anarchy; the inconveniences being all as great and as near, but the remedy farther off and more difficult.[5]

Because such great mistakes of government may occur in democracies or dictatorships, irregular wars do not require a certain political structure for the sovereign as a causal mechanism. It is the functional or dysfunctional sovereign that sets conditions for social revolution and irregular war. However, irregular wars are far more likely in authoritarian or failed states since democracies by design possess features capable of mitigating those conditions, lowering the probability of such conflicts. Figure 6 provides a visual representation of the irregular war condition arising from the interaction between the sovereign and the social order.

Sovereign dysfunction is an independent variable influencing the integrity of a social order. The ideally functional sovereign sustains the ideally integrated social order. As the sovereign descends deeper into dysfunction, the social order begins to disintegrate. A totally dysfunctional sovereign presiding over a totally disintegrated social order results in an irregular war condition that can potentially lead to total collapse of society and the state. An ideally functioning sovereign can help to rescue a disintegrating social order through the mechanisms of its positive functionality, such as the administration of justice to overcome the loss of socially

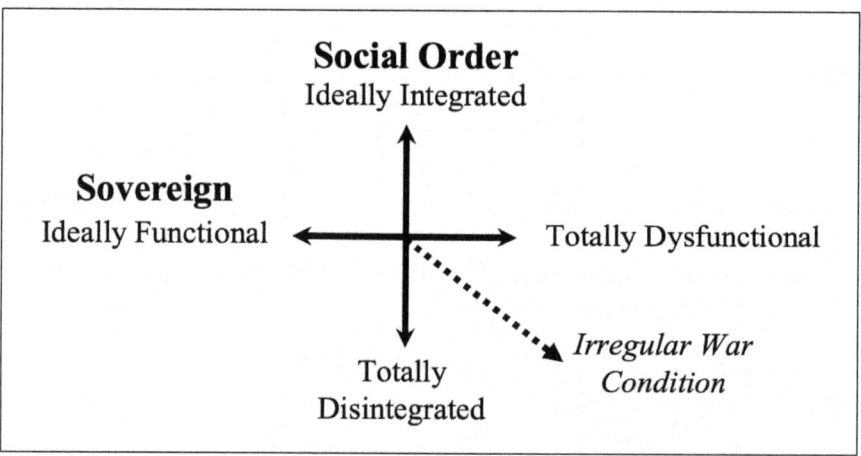

Figure 6. Sovereign Dysfunction and Social Order.

constructed moral restraints. The social order can be ideally integrated without a sovereign at all, as in the case of hunter gatherer groups. However, once power structures begin to take shape that outlive the participants, sovereignty has emerged, and its level of functionality will influence the social order for better or worse.

Like the structural element of the polity, certain scales or degrees of liberty do not establish necessary conditions for transitions between dictatorship and democracy. Instead, categorical and fundamental transformations of political power and structures of sovereignty must take place to change the social order.[6] For example, Brazilian revolutionary leader Carlos Marighella clearly defined the goals for his *Ação Libertadora Nacional* (National Liberation Action) in Brazil in these terms: "Dismantle and destroy the present ... economic, political, and social system" and establish "a totally new and revolutionary social and political structure, with the armed population in power."[7] Fragmented power-sharing schemes among the elite and a dictator may contribute to a successful uprising, especially when the state lacks parties and legislative systems sufficiently balancing popular and elite grievances to the benefit of the dictator.[8] Indeed, conflict among elites preceded regime change in authoritarian systems in an overwhelming majority of cases compared to insurgencies and revolutions from below. Of the 303 authoritarian leaders forced out of power against their will from 1946 to 2008, 68 percent left by coup, 11 percent left by popular uprising, 10 percent transitioned to democracy, 7 percent were assassinated, and 5 percent left after foreign intervention.[9] These transformations often create a liminal sovereignty in the ashes of contested claims to the legitimate use of force among violently competing factions.

Authoritarian states often organize around a kind of "romantic territorial nationalism,"[10] which such regimes simultaneously foster and feed upon at the exclusion of any competing narratives of power and identity over the same sovereign territory. Authoritarian governments themselves are often born out of a failed transition to democracy or during an independence movement. In fact, authoritarian governments formed since 1946 were a result of a breakdown during the process of transitioning to democracy 29.15 percent of the time, while 36.18 percent occurred during an independence movement. Further, 17.09 percent of these began out of a lack of sovereign authority in a territorial space.[11] Conversely, authoritarian governments ended 48.74 percent of the time due to transition to democracy, while only 19.1 percent ended due to lack of sovereign authority. This is true for 199 "authoritarian spells,"[12] which are periods of dictatorship, continuing uninterrupted for a time from 1946 to 2008. Still, whether successful or not, these political conditions often result in periods of irregular war due to the liminal sovereignty arising during unsanctioned power transitions.

Liminal sovereignty occurs at a moment of power transition, in which an incumbent is faced with a challenger defying an existing institution or process designed for the purpose of legitimate turnover of authority. Bernard Tricot, chief negotiator between France and the Algerians at Evian in 1961 asked how a "revolutionary organism which had created itself into a government without ever having exercised any territorial authority and pretended to represent a state which had never existed" could in fact achieve statehood.[13] A prominent historian of the French-Algerian war concisely explained this contradiction, noting that "in Algiers there was now no governor-general; in Paris, no government. A chasm gaped enticingly."[14] That gaping chasm was the lack of a functional sovereign expressed through the state.

States can be thought of in this sense as "coercion-wielding organizations that are distinct from households and kinship groups and exercise clear priority in some respects over all other organizations within substantial territories."[15] Their power structures also have the character of "perpetually-lived organizations."[16] In other words, states are meant to outlive their constituents. Sociologist Max Weber identified traditional, charismatic, and legal legitimations as three forms of authority states take to maintain their monopoly on force, which he termed "legitimations of domination."[17] One common view that follows Weber's definition of the forms of authority asserts that societies expect legitimate managed violence to be used in ways that the constituent society generally approves of.[18] But what is legitimacy? A useful definition for legitimacy in the context of people and their states is "the belief of the populace and the elites that rule is proper and valid, that the political world is as it should be."[19] When legitimacy is perceived to be lost, the state disintegrates, and a struggle begins over the errant power now unconstrained and with a force now illegitimate.

Following Weber, those viewing states as black boxes tend to emphasize the monopoly on the legitimate use of force as a mark of sovereignty within the state so that it may engage in interactions with other states.[20] This sovereignty is significantly contested once legitimate public agents of the state lose their monopoly on force, leading to liminal sovereignty and conditions for irregular war. Writing in 1930, Leon Trotsky saw the breakdown of social order occurring in the wake of "some swift growth which has broken down the old equilibrium of the nation."[21] These state-centered approaches to force and power hold that political violence results while established and emerging "selectorates" vie to produce a winning coalition capable of consensus.[22] In this view, the winning coalition can resume effective interest-protecting behavior within a political system rather than continuing the violence only if a functional sovereign is restored.

Chapter 7. Irregular War Conditions

The breakdown of the monopoly on violence has complex causal pathways, stemming from equally complex events like war, economics, politics, and even elite behavior or the environment. Regimes are prone to violent overthrow if their performance in a recently ended war is perceived to be poor.[23] From an economic perspective, a dysfunctional sovereign disrupts the rent equilibrium among and within states, thereby placing a greater incentive to arm and resort to violence because the safest assumption is that the others in disequilibrium could do the same.[24] However, elites prefer to protect an incumbent regardless of the degree of sovereign dysfunction if their interests are threatened by an alternative with an unclear outcome because of the sunk costs the elites have already invested in that sovereign. Prospect theory proposes that sunk costs and unknown benefits influence elites to rally around a dysfunctional sovereign in such a way that its effects are insulated from levers affecting social order. Aspiration and ideological unity are insufficient conditions to challenge the political system and its leaders, a claim observed consistently throughout the arc of political violence in Syria from 1982 to 2022.[25]

Institutions and public forums give elites and the masses pressure valves to divert grievances away from the sovereign, thereby contributing to stability even if those valves are artificial or marginally productive venues for change.[26] This "escape valve" is a spoiler for revolutionary social movements that are intentionally structured group actions seeking transformation, modification, or abolition of the existing system.[27] Leon Trotsky famously noted that "a revolution takes place only when there is no other way out."[28] This escape valve effect is not restricted to Western or European states, nor is the effect on the growth of the movement if the valve is not opened. These types of movements in the late Ottoman Empire and Sassanian period in Iran both demonstrated that one speaks with the sword only when the voice is silenced.[29] States with open access political-economic systems are inherently less internally unstable and less prone to internal violence because of the way organizations and rights are structured in these states.[30]

Further, richer states often have more organizations than poorer states, providing more escape valves for those agitating against the social order. These social orders control resources and social behaviors while mediating violence by controlling incentives and rights for certain members of the social order.[31] There are several distinct escape valves that have had regime-stabilizing effects at the level of the ruling elite, ameliorating opposition to even the most dysfunctional systems. These include resolving contentious policies, averting crises, presenting the facade of political transition, and keeping competitors sufficiently impotent as to pose no real threat to the incumbent.[32] Deng Xiao Ping employed a strategy with similar effects after succeeding Mao Zedong, powerfully stabilizing the sovereign while reducing the onset of irregular war conditions within the social order.

Circumstances enabling a transformation from aspiration to collective action must exist and be opportunistically manipulated to gather sufficient opposition to challenge the dysfunctional sovereign. Without these conditions, insurgencies will fail to exploit conditions necessary to change the social order. Russian history provides an important series of connected examples. Russia experienced four major insurgencies from 1600 to 1800, led by Bolotnikov, Razin, Bulavin, and Pugachev, respectively.[33] All four were fought against landlords for liberation from regional centralized authority rather than against the tsar. None succeeded in achieving their aims. The tsar's absolute power over the Russian Empire prevented the sufficient revolutionary moment favorable for collective action until after the duma was formed in 1905. The duma was insufficiently empowered as an escape valve for popular dissatisfaction with the dysfunctional sovereign, leading to an abortive revolution in February 1917. This long series of events, taken together, set conditions of liminal sovereignty necessary for the successful revolution in October 1917.

The causal pathways leading to irregular wars overturning social orders are extensive, often spanning generations. Arab nationalism grew from attempts to exploit sovereign dysfunction and transform the social order amid the liminal sovereignty of the waning Ottoman Empire and the growing Anglo-French involvement in that empire's withdrawal from the region. The first organization agitating for Arab independence was founded in Beirut in 1876 after the functional sovereign authority in Istanbul was disrupted with the Ottoman *tanzimat* reforms. Beginning in 1881, secret pamphlets circulated in Beirut calling for Arab independence from Ottoman rule. The *tanzimat* reforms led to rising nationalism throughout the Ottoman Empire after the Ottoman sultan opened the escape valve more quickly and broadly than the existing social structure could abide as he lowered himself from a supreme authority to what was ostensibly merely a party to a contract with his subjects.[34] In Lebanon, the reforms led to uncoordinated decentralization of sovereign power followed by the rise of a confessional authority allocation system expanded later under France.[35] Central political authority in Lebanon was impossible at the state level because communal and localized sovereignty were consolidated among sectarian elites rendering the state effectively impotent over Lebanese society and territory.[36] These and other events led to the *nahda* (renaissance) in Lebanon. The *nahda* was an intellectual movement that spread throughout the Arab world from 1860 to 1914, transforming thought into action.[37] During this time, other intervening

Chapter 7. Irregular War Conditions

conflicts accelerated the process, including the Russo-Turkish War (1877–88), the British occupation of Egypt in 1882 and Cyprus in 1878, and the French occupation of Algeria in 1830 and Tunis in 1878. These sentiments coalesced into a nascent Arab nationalism, setting the stage for irregular wars challenging the social orders during revolutionary moments to come in Cyprus, Lebanon, Algeria, Iraq, Syria, and Morocco, among other places.

In this revolutionary moment a social order is faced with competing sovereignties amidst a population sufficiently engaged for collective action.[38] The case of Iran in February 1979 is illustrative of the importance of the revolutionary moment, in which an "explosive social and political broth of long-simmering grievances began to bubble over"[39] leading up to the revolution. Iran's revolutionary moment in 1979 actualized those simmering grievances felt within the social order against the dysfunctional sovereign. That moment arose in the intersection of the Shah leaving the country, the opposition leadership arriving amid great fanfare, and primary groups in Iranian society seeking to change the existing social order.

The Armenian-Azerbaijani conflict (1905–06) is a lesser known but illustrative example of a failed attempt to overturn a social order despite the existence of sovereign dysfunction. An analysis of 245 letters written by the participants revealed four major causal factors contributing to the onset of the conflict and its failure.[40] First, the Armenian Dashnak party was employing a "despotic method of rule" over areas with mixed Orthodox and Muslim populations in the Caucasus, characterized by "bloody acts of terror."[41] Second, the political authority Russia traditionally exerted in the Caucasus was absent after the 1905 Russian Revolution and the subsequent Russian loss in the Russo-Japanese War (1905–06). Third, the lack of information and access to education left Azerbaijanis in particular without time to gather weapons for events they had no idea were about to occur. Fourth, the Armenian goal of autonomy or independence from Russian colonization and Ottoman sovereignty was rejected by both powers, leaving Armenian nationalism as a pawn in the festering territorial ambitions of the Ottoman and Russian empires still unresolved after the Russo-Turkish War (1877–78). The Armenian-Azerbaijani conflict (1905–06) lacked the necessary revolutionary moment for effective mobilization against the dysfunctional sovereign.

Although the people may be burdened with numerous grievances, these grievances cannot result in empowering collective action with some other element of the social order unless the grievances are perceived to be held in common and sufficiently profound to call factions to action.[42] There are several ways the public may decide to defect from political control and begin a violent transition from one form of government and structure of power to another.[43] The sovereign may convert from within, thereby retaining a modicum of legitimacy and appearing to monopolize the provision of public goods. The regime may open valves of expression in its attempt to accommodate demands, but this may instead serve as an open flood gate. The regime may be coerced out of power, even with non-violent means such as mass protests and physical occupation of government spaces as happened in Serbia in 2000. Finally, the regime may disintegrate and dissolve after legitimacy has already begun to evaporate, transfer, or transform. The sovereign depends upon the social order for its legitimacy no matter how authoritarian the regime or how despotically its power is wielded.

Conditions of Sovereign Territory and Sovereign Dysfunction

Sovereign territory consists of that physical land which is governed by the state and all the property, private or public, held among subjects to its authority. For John Locke, the sovereign existed to ensure life and liberty, but the protection of property was "the great and *chief end.*"[44] Nineteenth-century French economist Frédéric Bastiat argued that the sovereign had the opposite effect, perpetually plundering individual property.[45] These European conceptions of property contrast with non–Westphalian political entities, such as the Seminole Native Americans. The Seminole matrilineal clan system relied on clan councils to determine land use rights, which stood in opposition to English common law, the latter having heavily influenced American property law in the nineteenth century. The contradiction between communal and individual property rights caused friction between frontiersman of European descent and the Seminoles.[46] However, the centrality of territory to sovereignty cannot be ignored, regardless of the structure of the relationship between the two. Military theorists continue narrowly defining irregular wars as "contests where populations, rather than territory, are decisive,"[47] sidelining the effects of the relationship between sovereign territory and sovereign dysfunction in establishing a condition for irregular war.

A condition for irregular war exists when the sovereign fails to assure property can be protected or to adequately defend its own territory. A two-part test can be used to determine whether a state is sovereign in this regard.[48] Before setting out the rule of the test, the elements to be tested must be defined. These elements include sovereignty, territorial states, and the structure of legitimacy the state employs. Westphalian sovereignty is a preserve of territorial states and consists of a power "superior to the other forces which made themselves felt in that territory." Given this characteristic, it is important to determine whether a state actually possesses authoritative power over a territory and to ensure there is no other power superior to it.[49] The territorial state is an integral unit with conspicuous boundaries, and it contains "the administrative and coercive agencies possessing legitimate authority over a particular territorial order."[50] The first step of the test is to determine what degree of legal control other states exercise over the apparent sovereign. The second step is to determine which state or entity actually controls the functions and institutions inside the given state. This test can be applied in any irregular war case to identify where sovereignty actually lies and how institutional capacity is distributed among those contesting that sovereignty.

Most wars between or within states are fought over disputes of territory, regime type, or policy, with issues of territory being the most common by far.[51] War, regardless of form, therefore usually involves a dispute of territory or property within the greater conflict framework. Irregular wars may consist of violent territorial disputes between sovereign states and factions that are not states. Figure 7 depicts the irregular war condition arising when a totally dysfunctional sovereign presides over a totally uncontrolled sovereign territory, and the relationships among different arrangements of territory and sovereignty.

Any analysis of sovereign territory and war must occur within the context of the Westphalian national state and European dominance of international politics from 1648 forward. Europe controlled just seven percent of the total land area on earth in 1500, but this proportion increased to 84 percent at the start of the First World War.[52] State borders serve the primary military, legal, economic, ideological, and social-psychological

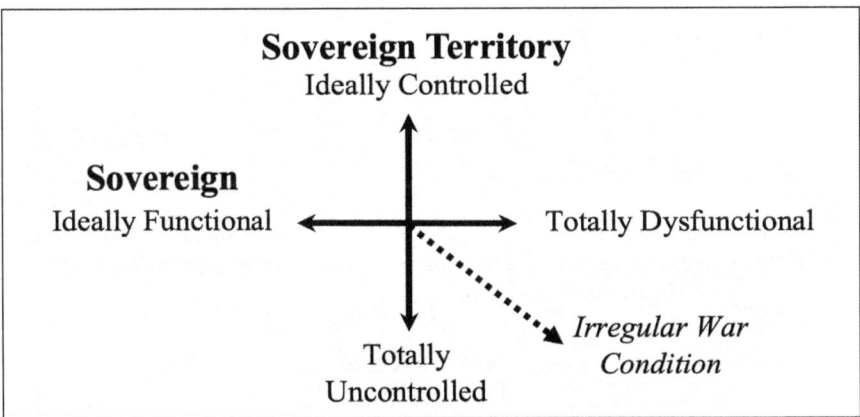

Figure 7. Sovereign Dysfunction and Sovereign Territory.

functions for the society or societies over which they exercise authority in the Westphalian system of state sovereignty.[53] From another perspective, there exist three "synonyms of sovereignty,"[54] or features of the territorial state, including political independence, equality of the sovereign in terms of other states, and unanimity of the sovereign in the weight of its vote in the international community against other states regardless of size or capacity. Measures of the state in terms of functions and features are helpful to describe the expected relationship between the state and its peers or its constituents, but the existence of irregular wars resulting from sovereign territory interacting with a dysfunctional sovereign suggests these heuristics are insufficient for a complete theory of state sovereignty.

The state as an independent political association exerts influence over its constituents and its peers with the power of force, whether implied or imposed. The concept of state formation is generally agreed to occur only through force.[55] Those quoting Max Weber's popular definition of the state as an entity seizing and maintaining "the monopoly of the legitimate use of physical force" often omit a crucial element of his concept of the state: legitimate violence must be exercised "within a given territory," thus linking physical sovereignty with a territorial state.[56] In other words, all power within the state is contested to the extent that it is legitimized and constrained by and for the state within this definition. Weber gave a second definition of the state as "a compulsory association which organizes domination," omitting physical territory as an element of this sovereignty.[57] This second definition comports with the notion that state and social boundaries can be coincident with one another.[58] In reality, however, those boundaries are not so neatly laid.

There are important variations in state sovereignty that may play a role in irregular wars which emerge from the interactions between sovereign dysfunction and sovereign territory. French anthropologist Claude Levi-Strauss described the "science of the concrete,"[59] or a deep desire among people to define their physical space. This desire gives meaning to otherwise meaningless objects and demarcates the physical boundaries between "our land" and "their land," without requiring that "they" accept "our" boundaries. Using such a negative test to determine whether a sovereign exercises power over a territory can result in misleadingly identifying irregulars primarily through their separation from existing state powers as though they resist sovereignty in general. This is misleading because most non-state actors in irregular wars seek the very sovereignty guaranteed to states within the Westphalian system.[60]

Conditions for irregular war may therefore arise when the power, force, and tangible exercise thereof become contested in a territory governed under a dysfunctional sovereign. There are many forms these conditions may take. A state may be an *imperium in imperio* (state within a state), such as Puntland, Azawad, or Hezbollah's governance of south Lebanon. The Cuban *territorio libre* (liberated territory) in the revolution (1953–58) is another example of a state within a state, albeit in a different conflict structure.[61] The process may occur outside the territory as well, as the Chinese Kuomintang set up a "provisional government/guerrilla army scheme" in Korea to function as a shadow state underground against Japan in 1942.[62] Chiang-Kai Shek advocated creating a similar underground apparatus in Thailand in 1943. Others include the Taliban shadow government based in Pakistan but governing in Afghanistan (2001–21), Da'esh in Syria and Iraq (2014–17), and the *Fuerzas Armadas Revolucionarias de Colombia* (Revolutionary Armed Forces of Colombia/FARC) in Colombia (1964–2017). Quasi states may exert some form of state-like territorial sovereignty, such as Rojava, Somaliland, and Artsakh. Semi-autonomous states including Kosovo, Taiwan, and Banda Aceh exert territorial sovereignty with the conditioned consent of a Westphalian state that is challenged significantly by others. Nominal states such as Palestine and Western Sahara exert limited sovereignty over a territory due to military occupation from a neighboring state. The Casamance and Guayana Esequiba are expressions of a violent territorial dispute, the former internal and the latter between states. South Ossetia, Abkhazia, Northern Cyprus, and Transnistria are partially recognized states existing because a strong Westphalian patron shields them from dissolution. However, sovereignty in an abnormal form does not *io ipso* invite violence, let alone irregular war.

There are many examples of sovereign or semi-sovereign power arrangements in the Westphalian system that are relatively stable due to a functional sovereign executing authority over a territory. Belgium and the Netherlands provide one interesting example in the town of Baarle-Hertog, in which two states execute peaceful, consensual dual governance over the same territory. A less formal example of this exists on the German-Danish border, comprised of North Schleswig and South Schleswig, in which the former Duchy of Schleswig is split between the two states. This immediately calls into question political scientist Hans Morgenthau's claim that "sovereignty over the same territory cannot reside simultaneously in two different authorities; that is, sovereignty is indivisible."[63] Statelets like Vatican City, Liechtenstein, San Marino, and Andorra depend upon their neighbors for provision of certain services typically expected of an independent state including security, economic, and other functions. The United Kingdom's continued administration of Gibraltar after the Treaty of Utrecht in 1715 demonstrates an example of peaceful extraterritorial sovereignty despite being contested by Spain. The case of Aurocania-Patagonia is another example of a non-violent declaration of sovereignty over territory, even enjoying protective diplomatic status and nominal consulship in Mogador, Morocco in the 1880s.[64] The Spanish exclaves of Ceuta and Melilla have achieved relative stability in northern Morocco, while some overseas departments such as French Guiana and Mayotte exist somewhat peacefully with their sovereign neighbors, as do the constituent states of the Dutch Caribbean.

States are expected to control territory, administer justice, maintain a legal system, extract financial resources from the population, enforce the law, and so on. Da'esh is a useful example to argue for at least one necessary condition for statehood: recognition from at least one other state in the international system.[65] Without this, sovereignty remains purely liminal if not imaginary. From 2014 to 2017, Da'esh certainly controlled contiguous territory across Syria and Iraq, administered justice through *hisbah* police powers, maintained a complex legal system based on an interpretation of sharia law, collected various taxes through several methods, and executed many other state-like activities. However, without recognition from just one state, Da'esh was essentially a high-capacity armed group monopolizing violence in its territorial environs. Da'esh competed with governments in Syria, Iraq, Afghanistan, Philippines, Mali, and other places for the undisputed sovereignty over territories within those states. The competition was lost at the outset because no sovereign state accepted Da'esh as a state. A widely accepted view originating after the Second World War held that there could not be competing sovereignties, but this argument applied to states and their relationship to higher institutions, such as the League of Nations or United Nations, not to states or other actors having competing claims to the same people or territory.[66] Conditions for statehood in the Westphalian system may neither guarantee *de facto* sovereignty nor do they prevent conditions for violence.

These conditions for violence arise from disputes over land, economics, and power structures. Disputes grow or whither depending on the level of cohesion, motivation to rebel, and grievance sufficient to push a social movement to organize and commit violence against the state.[67] Causes for fragmentation within political systems and fragmentation within the resulting factions both contribute to the potential for violence or conflict resolution.[68] Cohesion or fragmentation are affected most by repressive governments, accommodation from governments, the number of issues the movement addresses, and the presence of violence. There is a causal link between the level of cohesion in each group and the ability for such a group to mobilize units of collective action. Besides the use and extraction of resources, the physical terrain itself may serve as an impediment to cohesion.

Chapter 7. Irregular War Conditions 139

Geographic features can prevent sovereign states from establishing reliable, integrated lines of communication, logistics, and supply over the claimed sovereign territory. The Druze did exactly that, maintaining a modicum of sovereignty over their mountain redoubts in Lebanon and Syria during the Lebanese Civil War (1975–90).[69] The same features can inhibit the unity of the state, leading to competing concepts of belonging. Physical territory influences a "special geographic dialect" shaped by geography and social organizations, working together to bar far-flung urban elites from political meddling.[70] Sudan, the size of Texas, has about 400 languages and 600 tribes distributed among 597 ethnic groups.[71] Areas in Sudan are exceptionally isolated from one another, contributing to a feeling of absolute remoteness in a majority of the territory throughout the country.[72] Diversity was a long-term effect of this inaccessibility, allowing identities to develop in isolation while their interests diverged from one another, increasing the propensity to resist centralized authority of any kind from Khartoum.

Similar conditions for irregular war arose in Mali in 2012 after the government collapsed into a competition over territory in the absence of a functional sovereign, most notably in Azawad. The *Mouvement national de libération de l'Azawad* (Movement for the Liberation of Azawad/MNLA), an ethnic Tuareg confederation in northern Mali, declared the separate state of Azawad over the northern two thirds of Mali in 2012, comprising most of the Tuareg areas of the state.[73] The Tuaregs had long sought independence by staging rebellions, protests, and secession activities from Bamako since French rule forced ethnic Songhai hegemony over the Tuaregs in the late nineteenth century. France stymied Tuareg independence efforts, but the Tuaregs continued seeking autonomy, with the MNLA only the most recent expression of that goal.[74] Other groups in French North Africa have followed similar patterns.

The Algerian civil war in the 1990s produced several competing groups who sought territorial sovereignty and control of the political-economic institutions within those disputed territories. Notably, Hassan Hattab formed a Salafist group to implement Islamic sharia law, which merged in 2006 with the *Mouvement pour l'unicité et le jihad en Afrique de l'Ouest* (Movement for Oneness and Jihad in West Africa/MUJAO) and others to form al-Qaeda in the Islamic Maghreb (AQIM).[75] The secular and local MNLA competed with the Salafist and local MUJAO, while regionally focused Ansar al-Din and the Salafist AQIM also entered the conflict. These groups all drew ultimately from ethnic Tuareg agitations against the weakened state of Mali and related regional issues from neighboring states. Their cohesion was largely tied to the similar ideologies leaders of these groups used to frame their struggle. While the Tauregs in northern Mali and southern Algeria rebelled against the dysfunctional sovereigns in Bamako and Algiers, a different group stirred closer to the urban core. Soldiers in Bamako staged a *coup d'état* overthrowing President Touré in 2012, removing any semblance of a centralized functional sovereign. Meanwhile, MUJAO and AQIM contested the MNLA in Azawad, taking over Timbuktu along with some other Tuareg groups before Da'esh backed a local affiliate there.

In April 2013, the United Nations began its Multidimensional Integrated Stabilization Mission in Mali to provide an international response to the situation. Four months prior, France had launched its own counterterrorism mission, Operation Serval, to carry out its familiar strategy of *tache d'huile* and search and destroy, this time with United Nations approval.[76] Both missions attempted to stabilize the internationally recognized central government in Bamako despite its geographical isolation from the conflict zone in Azawad, rendering them mostly ineffective. Such a geographic dialect does not always require mountains, deserts, bodies of water or other physical terrain to create islands of resistance and sanctuary.

Central authority in urban centers is just as important as outside the cities. In the nineteenth century, Louis Napoleon directed Baron Haussmann to completely renovate Paris in large part to prevent insurgencies and revolutions from taking advantage of the difficult urban terrain. Indeed, from 1826 to 1861 insurgents had barricaded Paris nine times, including the longest insurgency of the century, which was the resistance movement against the coup that brought Louis Napoleon to power. That was until it was outdone by the Paris Commune of 1871 that sought to take the city back from the monarchy once more.[77] More recently, the *Front de libération nationale* (National Liberation Front/FLN) in Algeria used the urban jungle of the Casbah to resist the French while the unique structures in Iran's densely urban *bazaari* community helped shield those who overthrew the Shah in 1979.[78]

Geographic impediments to centralizing authority can be overcome with highly extractive economies such as those in Russia and China or through moderately inclusive political and economic institutions. In these cases, control is more powerful than geographic or local resistance though it may be unsustainable in the long term. Populations opposing overlords but strongly controlled by them offer muted or non-threatening resistance, making armed conflict less likely in the near term.[79] Central government must therefore be sufficiently tangible, whether in physical proximity or felt political effects, such as law enforcement, to be accessible and blameworthy among the people while offering nominal redress regardless of its effectiveness.[80] States lacking centralized control over their territory are more susceptible to the challenges to sovereignty promoting irregular war conditions. Extending authority through repression generates ephemeral control but the long-term effects of those policies have paradoxical outcomes as they lack the inclusivity necessary to contain, address, and resolve grievances in the state system, giving the aggrieved no choice but to subvert, avoid, or destroy that system.[81]

The state may reclaim its sovereignty over areas previously subject to liminal sovereignty or an insurgent faction can fill this void. India dramatically increased the number of police in the secessionist Andhra Pradesh while also developing infrastructure and introducing central authority through local bureaucracies, leading to a success like that of Magsaysay in the Philippines against the Huks.[82] From the opposite perspective, Algerian insurgents exercised sovereignty in certain areas during the Algerian Civil War (1996–99) and that relative peace prevailed in those places compared to contested areas.[83] This lends credence to Hans Morgenthau's claim that "the location of sovereignty may be in temporary suspense if the actual distribution of power within a territory remains unsettled."[84] However, Morgenthau's statement only holds true if either an insurgent force or the state can become the functioning sovereign in the previously contested space. One way this contest may be resolved is through state formation.

A state can form in a liminal space in many ways, whether through elections, conquest, or agreement. The Islamic Emirate of Kunar was established in Kunar Province, Afghanistan in 1990 after a local *shura* council declared *bay'ah* (allegiance) to Jamil al-Rahman, thereby legitimating just one state from among the existing "autonomous 'village states.'"[85] This structure of power lasted until it was destroyed by the more powerful regional Salafist leader Gulbuddin Hekmatyar with Pakistan's assistance.[86] States can form out of wars if there is sufficient growth in financial, security, and governing capacity in the nascent state.[87] Despite this, not all factions seek statehood in their violent conflict with the state.

Some insurgencies seek autonomy in ways that may not comport with the Westphalian system. Over three centuries, the Yazidis in Sinjar, Iraq rebelled against the Ottoman Empire, Britain, France, and subsequent iterations of the Iraqi government for this goal.[88] They sought neither statehood nor political office in the central government, however, nor did they seek financial concessions or unification of all Yazidi tribes in Iraq. This set of desired outcomes stand in stark contrast to the goals of the neighboring Kurds who sought all of these.[89] For Ansar al-Din in Mali in 2012, legitimacy flowed not from the distant and dysfunctional Malian government, but instead from the tenets of Salafism in the semi-autonomous Azawad area.[90]

In Iran in the 1980s, a battle was taking place not over territory but ideology. Ayatollah Hussein-Ali Montazeri justified foreign intervention and Islamic revolutionary internationalism in the Middle East in direct opposition to Westphalian conceptions of state sovereignty when he declared "Geography does not exist" in the theory of *velayat-e faqih* (governance of the jurists) promoted by Iran's clerical establishment.[91] Ayatollah Khomeini implemented *velayat-e faqih* in revolutionary Iran, a system in which the state governs by *fiqh* (jurisprudence) through *faqih* (guardians of law) interpreting Islamic sharia law with the assistance of *nezam* (jurists) under the Shi'a Ja'farī *madhhab* (school) of Islam.[92] Khomeini articulated the elements of this system in his 1969 manifesto *al-Hokamat-e Islami* while in exile in Najaf.[93] Like Montazeri, Khomeini framed the Iran-Iraq War (1980–88) as "Not a war for territory, it is a war between Islam and blasphemy."[94] The latter case contrasts the *ummah* (global Muslim community) against the secular order among peers in the Westphalian system. These examples illustrate a tension between the Westphalian system and other forms of sovereignty, in which the former places primacy on physical boundaries, while others may be temporal, spiritual, juridical, or familial. Regardless of the system of sovereignty, a violent contest over sovereign territory often creates a liminal space for a particular kind of war.

Irregular war conditions arising through contests of sovereign territory in the absence of a functional sovereign tend to manifest through insurrections and civil wars. In a 1932 letter to Albert Einstein, Sigmund Freud described insurrection and civil war as "a period when law is in abeyance and force once more the arbiter, followed by a new regime of law." In such situations, violence is more pronounced in areas of liminal sovereignty than in areas under complete control or on contested front lines, which represents an elemental difference between conventional and irregular wars.[95] The highest levels of violence in civil wars exist in situations in which control of sovereign territory is constantly in flux, while low levels of violence exist in areas of stable control regardless of the actor exercising such control.[96] An important example of the merits of this claim can be seen in the Nagorno-Karabakh conflict, which is defined by dramatic topography, accessibility issues, and separation from urban elites in Baku and Yerevan.[97] The war between Eritrea and Ethiopia (1998–2000) is another example of sovereign territory and sovereign dysfunction creating conditions for irregular war. This conflict was closely tied to the ambiguity of the border between the two countries after Eritrea gained independence in 1993.[98] In such a liminal space, contending forces vie for dominance over a sovereign territory in an irregular war.

The sovereign state is the dominant form of sovereign authority over physical territory in the Westphalian system. The dysfunctional sovereign may create conditions for irregular war when failing to effectively govern territory, exert authority beyond a central base, or ensure that external forces respect state boundaries. Not all situations of abnormal sovereign power arrangements over territory are susceptible to irregular wars since the variable of the sovereign and its functionality becomes the lever by which a state turns to dissolution, devolution, or disintegration as places including San Marino and Ceuta prove. The cases of Mali and Sudan also stand in contrast to Baarle-Hertog and Andhra Pradesh because of the position of this lever. Like social orders and political-economic institutions, territorial sovereignty varies based on its structure *vis-à-vis* the sovereign power of the state.

Conditions of Political-Economic Institutions and Sovereign Dysfunction

States in which capital is concentrated in finance are susceptible to democracy while states whose capital is primarily held in property resist it. Propertied power structures prefer hereditary government, nepotism, and unofficial systems of favors, among other preferences, while contributing to antagonistic power relations between sovereign and subject. Inclusive political and economic institutions combined with guaranteed rights for property, whether intellectual or material, allow a society to accept liberal democracy and its attendant benefits of prosperity, health, and freedom.[99] Liberal democracy in this context means a government with mechanisms for routine, constitutional ways of replacing top officials and having certain features including a political formula expressed through institutions, elected leaders, and competitors who can legitimately challenge the incumbents.[100] The effectiveness of institutions can be measured in terms of their value and utility in serving collective interests.[101] These democratic forms of political-economic institutions represent a functional sovereign, and those states having such institutions are unlikely to host irregular wars.

Dictatorship, democracy, and social revolution can emerge from economic conditions, social structures, and the political institutions developing to manage wealth and people.[102] Systems of inequality across social groups within a state contribute to the efficacy of repression through the military instrument of national power, setting conditions for the sovereign to function poorly amid a system of popular grievances related to hardship and poverty.[103] People have three options for dealing with institutions they dislike: they can attempt to change the institution through an existing legal process, they can vote with their feet by emigration, or they can

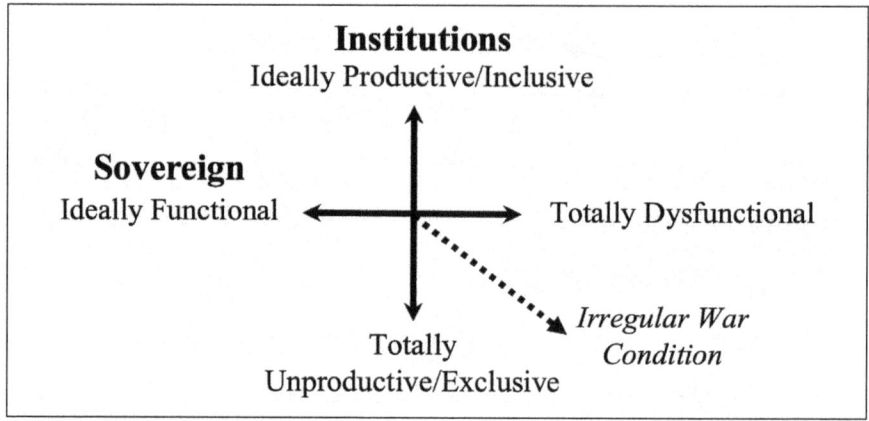

Figure 8. Sovereign Dysfunction and Institutions.

remain in place but use extralegal methods to force change, particularly organized violence.[104] However, the state's use of violent repression is a far more powerful instigator, incubator, and promoter of violent insurgency than competitive elections are at ameliorating, coopting, or eliminating them.[105] Figure 8 illustrates the irregular war condition that may appear when a totally dysfunctional sovereign administers totally unproductive and exclusive political-economic institutions, along with other arrangements of how the sovereign may interact with the political-economic institutions of the state.

Violent resistance to repression often manifests as insurgency and revolution directed at the obliteration of the dysfunctional institutions and systems of government so that the social movement can attempt to build new institutions and government on the ashes.[106] The new course may manifest as an attempt at state formation through the establishment of political-economic institutions carrying out the expected functions of the sovereign. Irregular war often occurs in the context of those conditions working to change a system of government by force through mass movements and institutional transformation to generate new social orders and political systems by force.[107]

The current structure of political-economic institutions in the Westphalian system developed in earnest around the seventeenth century. This structure has contributed to violent conflict in several ways when interacting with a dysfunctional sovereign. The period between the English Statute of Monopolies in 1623 and the publication of Adam Smith's *Wealth of Nations* in 1776 saw numerous developments emerge from the Industrial Revolution that led to the global spread and local adaptation of the political and economic institutions that are currently dominant.[108] These institutions grew out of a bargaining process between powerful factions in Europe who were not strong enough to replace the government but were still able to constrain the sovereign through their own representative participation in the political system.[109] These new representative institutions created novel power dynamics that changed the relationship between the government and the governed, constraining both in ways alien to the *ancien régimes*.[110] The resulting representative political-economic institutions shaped state behavior and affected societies in those states according to the interaction between the way the people expected the sovereign to function and the way it actually did function.[111]

The dysfunctional sovereign may interfere with political institutions that set conditions for the people to seek new structures outside a peaceful political process. The rigged election in Kashmir in 1987 motivated the resistance to India's central government, instigating separatist leader Yasin Malik to rally those disaffected with a failed political process.[112] Malik did this following an election event during which he witnessed a rigged ballot-casting. It was after seeing the failure of the political institutions to live up to the expected level of functionality that Malik decided to eschew rallies for rifles.[113] However, the issues leading to the uprising were long in the making, especially after the central government's decision to replace popular leader Sheikh Abdullah with an unpopular political appointee. This was accomplished outside the political process and prompted Pakistan to send more Islamist militias into Kashmir to attack central government interests there. Similarly, the insurgency in Argentina materialized after multiple groups suffering from various grievances coalesced into factions that were unsatisfied with the regime installed after the 1966 coup, especially as these grievances remained unaddressed through 1971.[114] The cases of Kashmir and Argentina are but two of many in which the sovereign interfered with the political process, prompting that fateful swap of ballots for bullets.

The dysfunctional sovereign may also manifest through extraction, inhibition, interference, or some other form of exploitation in the economy. Revolutions in Latin America were in part a violent rejection of the banana economy, highlighting the way extractive political-economic institutions directly contributed to insurgencies and revolutions in places such as Honduras, Guatemala, Nicaragua, and Panama, among others.[115] Those violent critiques of the dysfunctional sovereign are not limited to Latin America. The Dhofar Rebellion (1962–74) in Oman had many causal factors. However, the core issue of excessive taxation from the central government in Muscat drove the people in Dhofar to engage in an irregular war after finding no other recourse to address the tax policy.[116] Westphalian political-economic institutions at the level of the central government are not the only factors in such conflicts, as poor administration of tribal areas may also contribute to conditions for irregular wars.

In the case of tribal areas, a case of raids during the three Seminole Wars (1816–58) in the U.S. provides an example of a sovereign contributing to insurgency through dysfunctional political-economic administration of those lands. The U.S. government unilaterally cut rations promised in a treaty signed 18 September 1823 by nearly half, including for the staples of cattle and potatoes. Subsequently, Seminole raids on settler food stores increased as winter approached. The primary objective of these Seminole raids was to obtain food, with one such raid producing "some hogs, about 80 bushels of corn, and one kettle."[117] Rather than directly addressing the food shortage, Florida governor William Pope Duval requested federal troops to act as a deterrent force while keeping Seminole rations below the levels in the treaty and allowing "Poor conditions of the Native Americans and the constant intrusion into their lands by Whites hunting cattle and escaped slaves."[118] Later, a drought and unseasonably low water levels in the Ohio River affected both American settlers and the Seminoles, preventing imports from reaching the area to make up for the losses to both sides between the First and Second Seminole Wars from 1821 to 1835.[119] The Seminoles, by now emaciated and under threat of violence from the U.S. government, had no way to alter the dire situation through existing institutions. They had to choose between starvation or war. This case is but one of many similar instances of a dysfunctional sovereign interfering with expected outcomes of political-economic institutions often revolving around the twin issues of food and safety.

The needs of food and safety are indeed two recurring themes among those challenging the incumbent sovereign in irregular wars. Nineteenth century Russian aristocrat-turned-anarchist Peter Kropotkin identified poverty and militarism in the Russian political-economic institutions as the main forms of oppression requiring the overthrow of the state.[120] During the French intervention in Mali in 2012, one Malian academic noted, "We care about food, not democracy."[121] In Libya in 1997, Sheikh Subhi al-Tufayli led the *thawrat al-jiya* (revolt of the hungry) to protest against "poverty, poor health conditions, unemployment and government neglect."[122] This dysfunction in providing for the public good is not simply a matter of politics or economics; it may also affect the execution of justice, encouraging alternative forms of government to challenge the incumbent and install inclusive institutions.

Armed factions often make their aims plain as they embark on irregular wars. Musa Sadr, founder of the Amal movement in Lebanon, clearly outlined the purpose of that new organization in 1974: "We are rejectionists, avengers, a people in revolt against injustice."[123] The Irish Sinn Fein movement is a secular example of the same set of goals. It sought to replace British rule with its own republican government, police, and judicial systems in Ireland in 1919.[124] In the Marxist-Leninist context, Che Guevara made it clear that the Cuban revolution sought to target and replace the political-economic institutions of the state. In other words, victory meant that "the regime has been systematically and totally smashed."[125] In North Africa, the FLN outlined their goals on a radio broadcast into Algiers from Cairo on the first day of the war on 1 November 1954: "After decades of struggle, the National Movement reached its final phase of fulfillment.... Goal: National independence through restoration of the Algerian state, sovereign, democratic, and social, within the framework of the principles of Islam," while invoking the United Nations Charter along with a demand for "recognition of Algerian sovereignty."[126] In Cuba, Che Guevara asked rhetorically "Why does the guerrilla fighter fight?" His own answer was that "the guerrilla fighter is a social reformer, that he takes up arms ... and that he fights in order to change the social system that keeps all his unarmed brothers in the conditions of the reigning institutions at a particular moment and dedicates himself ... to breaking the mold of these institutions."[127] The process of rejecting the political-economic institutions of the state may also take the form of resistance to a new system over one already in place.

The Yazidis in northwestern Iraq organized resistance to a British colonial policy designed to destroy their traditional social and political institutions in the nineteenth and early twentieth centuries. Like the Yazidis, the Druze in Syria and Lebanon lived under their own *de facto* traditional authority structures in the eighteenth and nineteenth centuries in defiance of the Sublime Porte in Istanbul.[128] The Yazidi leader Dawude Dawud led the Yazidi resistance against British attempts to impose political-economic institutions on traditionally governed Yazidi areas in the 1920s.[129] French colonial authorities based in Syria attempted to take advantage of this by offering money to the Yazidis in exchange for ceding Sinjar in present-day Iraq to French-controlled Syria.[130] The Yazidis violently resisted both the British and French attempts to interfere in their institutions.

Insurgents may develop "counter-states" in areas of liminal sovereignty using parallel systems of taxation, government services, and security mechanisms to replace or divert the dysfunctional political-economic institutions under the incumbent. For example, the *imperium in imperio* in western Ukraine under the Ukrainian Insurgent Army in the 1950s closely resembled the political-economic institutions Hezbollah employed during the late 1980s in the Lebanese civil war and afterward.[131] Communist guerrillas in Greece had transformed quickly in 1943 "from a clandestine organization into nothing less than a state"[132] through their coercive arm to issue official documents, enforce laws, and administer justice.[133] This coercion was conducted with the expectation held within the insurgency that the coercion was legitimate in the same way a state monopolizes coercion. More shrewdly, Vladimir Lenin proclaimed that "the fundamental question of every revolution is power."[134] This power is exercised through the coercive activity of the state or its competitor. Joseph Stalin went further: "The whole point is to retain power, to consolidate it, to make it invincible."[135] State power without those characteristics is open to contest. In an otherwise different context, the competing legal systems of *pancasila* (five ideals) and *fiqh* (Islamic jurisprudence) in Bandah Aceh and elsewhere in Indonesia led to insurgencies as the sovereign contended with rival sources of law and justice. Those two legal systems are incommensurable: *fiqh* and Islamic sharia law support the establishment of an Islamic state in portions of Indonesia, while *pancasila* provides for a secular state built on nationalist principles.[136]

Insurgents may implement institutions in enemy-controlled territory to establish the facts of the state when the enemy sovereign cannot function adequately, as happened in South Vietnam in the 1950s and 1960s. North Vietnamese efforts to establish their own legal, financial, and civil services in place of the incumbent in South Vietnam created competing systems that both had merit in their viability.[137] North Vietnam's strategy began materializing as early as 1957–59 and the objective was to isolate the U.S.-backed authoritarian, centralized power in Saigon away from the outlying 17,000 hamlets and 8,000 villages in South Vietnam's countryside.[138] The North Vietnamese then established schools, hospitals, and public works in the south between 1961 and 1962, running a fully operational census department, farm bureaus, tax collection services, and news agencies there by 1963–65 despite a peak of 536,000 American forces occupying the urban centers by 1968.[139]

Similarly, Cuban revolutionaries sought to establish industries and create the trappings of the state capable of extracting taxes during the Cuban revolution.[140] Civil non-combatant organizations were vital for supporting the insurgency, and the revolutionaries took full advantage of this.[141] Foremost among these was "the council," a government in miniature that was organized to administer justice, execute laws, and preside over public life.[142] These elements were notably absent in Cuba before the revolution due to the lost legitimacy of the collapsing U.S.-backed Batista state. This strategy was not limited to Marxist-Leninist movements in places such as Cuba and Vietnam.

The Indonesian National Revolution (1945–49) employed the protracted methodology common to Marxist-Leninist movements, implemented instead by a secular Javanese insurgency seeking republican government, the right to vote, and independence from a foreign power.[143] The territorial organization of the Indonesian resistance to the Dutch in Java was composed of a local political and security apparatus that provided material support, shelter, and information so effectively that the insurgents needed only their clothes while moving from village to village.[144] During the uprising on Java, a member of the Javanese resistance observed that "a protracted and total people's war ... is a domestic military-political-economic problem, which we must tackle with all the energy and strength at our disposal."[145] The Javanese insurgents framed their violent contest with the Dutch in terms of a battle over the instruments of national power rather than territory or groups of people.

Irregular war conditions may arise from the political-economic institutions affected by the activities of a dysfunctional sovereign. The sovereign may circumvent the political process, suspend rights, inhibit justice, interfere with traditional institutions, or support an extractive economy. The people may reply with insurgency to remove this dysfunctional sovereign and establish political-economic institutions that reliably produce expected outcomes. The violent resistance to those extractive or undemocratic institutions may come in the form of a rival sovereign, competing government functions, or establishment of services in areas otherwise outside the control of the central authority. These and other conditions contribute to irregular wars, and the people who fight them will next be considered.

CHAPTER 8

The Elements: People, Politics, and Propaganda

> How many things apparently impossible have nevertheless been performed by resolute men who had no alternative but death?—NAPOLEON BONAPARTE[1]

Thinkers and Actors

The people are the insurgency. These people consist of important segments of society each serving certain necessary roles led by influential thinkers and actors. Civil society contributes thought leaders, collaborators, supporters, and passive resisters. Religious organizations, political parties, and trade unions are three powerful civil society elements capable of mobilizing coercive force against a government, an opposition group, or an external competitor.[2] Messaging favorable to the movement is also necessary to shape, direct, and operationalize public attitudes toward the ends the group seeks.

The Javanese insurgency in the Indonesian National Revolution (1945–48) is an illustrative example of civil society performing multiple functions for the Javanese resistance to the Dutch. The traditional *lurah* (village headmen) retained legitimacy and power while the insurgency educated and recruited from among the *pemuda* (youth) in those villages. The insurgents then acquired funds in the form of a "supervisory donation" with the help of the *lurah* in each area.[3] The burden of resistance was distributed across villages, towns, and ethnic groups to ensure all carried some responsibility for the conflict and for the future of Java.

The popular support for the insurgency from elements of the public can be expressed in terms of the latter's affect and duration of support. The supportive affect can be depicted on a continuum from sympathetic to fearful, while the duration can range from sustained to incidental. These

Chapter 8. The Elements: People, Politics, and Propaganda 151

civilian attitudinal conditions are represented visually in Figure 9.[4] The least invested support to the insurgency is uncommitted compliance, whereby a person provides acute support out of fear of reprisal, such as hiding an insurgent in a closet as a military patrol sweeps the area. Slightly more invested, people expressing habitual compliance do so over long periods of time and, like uncommitted compliance, fear is the motivator on their affect. Examples of habitual compliance include paying taxes to Da'esh members controlling Mosul from 2014 to 2017. Crossing a boundary in affect from fear to sympathy changes the structure of support that civilians will provide to the insurgency. The lesser of these is contingent support, which occurs when a person is sympathetic to the insurgent's cause but only offers acute assistance. The network of civilians across Nazi-occupied France who offered their homes to the French *maquis* for temporary shelter during the Second World War is an example of contingent support. Finally, the most committed support to the insurgency arrives in the form of sympathetic support, which occurs over a long duration and is consistently voluntary.

The insurgency itself consists of political-military leaders, support personnel, and operational forces. The state contributes security and military forces in opposition to the insurgency, while third-party interventions often consist of some form of the same whether in support of the insurgency, the state, or both.[5] Security and military forces are separated along lines legally demarcated by statute and in terms of the needs of national security.[6] Security forces could include any element conducting covert action, sabotage, subversion, espionage, and other special or intelligence activities supporting national security objectives. Military forces include those conducting traditional military activities such as conventional maneuver, terminal attack control, unconventional warfare, special reconnaissance, direct action, security cooperation or assistance, and other operations, actions, and activities oriented toward military objectives. This distinction is between authority and purpose rather than merely being a matter of definition. Individuals can flow between all these roles depending on the situational pressures and group dynamics in context.

To achieve results, the insurgency depends on cohesion, group identity, and obedience. The most important factor for the long-term success of an insurgency is sustained cohesion in the face of external or internal pressures to dissolve the group. To achieve this, the insurgency must generate a group identity more powerful than the individual so that the individual surrenders some of their liberty in exchange for the perceived benefits of supporting the insurgency. There are three processes contributing to the formation of a group identity: *inclusion*, becoming an insider; *collectivism*, thinking how to contribute to the group rather than simply

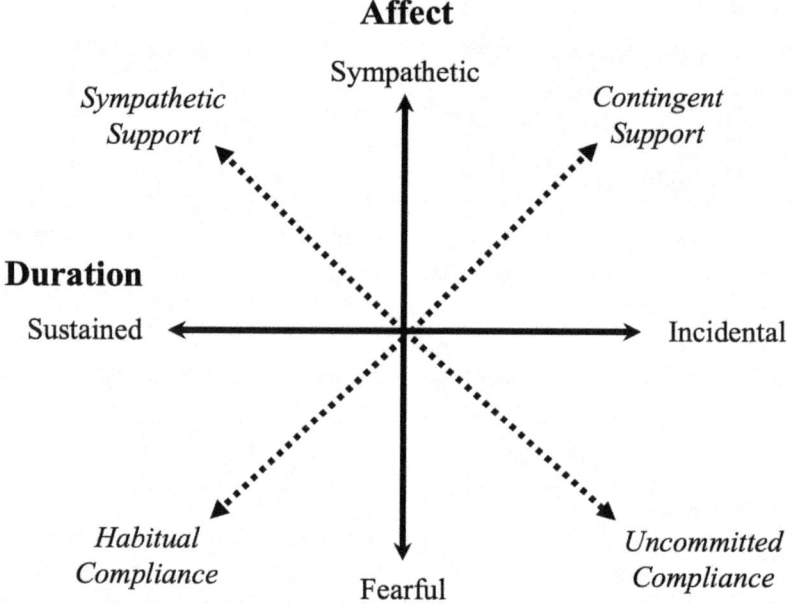

Figure 9. Civilian Attitudinal Conditions Supporting Insurgency.

benefiting from inclusion; and *identity*, reifying the group identity with the individual member's identity.[7]

The process of group identity formation is not unlike the surrender of liberty for obedience in social contract theories of the state. Thomas Hobbes argued that citizens exchange obedience for life and liberty, a trade he described as "the mutual relation between protection and obedience."[8] It follows that political power is predicated on obedience.[9] People obey out of habit, fear, self-interest, moral obligation, affinity to the ruling class, indifference, and lack of confidence. Political power absolutely depends upon obedience.[10] It is the threat of force such power credibly conveys that invokes obedience. It is unsurprising that PKK leader Abdullah Öcalan framed the force immanent to political power as the expression of state violence when leadership fails.[11] It is political power that insurgencies seek, and an element of sovereignty is the accumulation of sufficient obedience rendered to justify the application of force. This political element comes in the form of legitimacy, which is one of the six sources of power.[12] Like a conventional military, irregular forces need to successfully manage recruitment, retention, and desertion.[13] The insurgency must develop strategies to attract potentially obedient recruits.

Chapter 8. The Elements: People, Politics, and Propaganda 153

Insurgent recruitment benefits can be expressed in terms of the perceived value disparity between the way things are without the insurgency and the way things are promised to be by the insurgency leadership. This benefit is further tempered through factors including group credibility, ideology, and affinity between the pool of potential insurgents and the insurgency leadership.[14] The collective action problem places a rational constraint on the social acceptance of insurgency and revolution through the costs imposed on the individual recruit or member without guarantees of either success or reward sufficient to justify the costs.[15] However, insurgents are simultaneously attracted to organized violence against the state by the magnetism of primary group cohesion rather than just ideology, higher morals, politics, "true belief," or doctrinal commitment.[16] The nature of insurgency and revolution requires that a shared expression of organized violence is defined by its contrast to the dysfunctional sovereign.[17]

This characteristic of an insurgency cohering around opposition to an incumbent sovereign has been explored many times, though the specific connection to the effects of a dysfunctional sovereign upon the coherence of the insurgency has not. Persecution is a powerful condition contributing to building group cohesion, forming a bond strengthened when external forces threaten the group as forces outside the group galvanize the unity of that group.[18] Some of the members of insurgencies fighting in twentieth century Syria calculated that they had a high chance of death whether they fought against the regime or not, so it was in their interest to die fighting.[19] Groups of individuals form the kernel of action during violent risings against the state, as each is empowered by the sense of deprivation the dysfunctional sovereign foists upon them.[20] Sometimes people are drawn to an insurgency not for the cause to which the leaders rally, but simply out of the desire for *gloire* (personal renown) or *thymos* ("spiritedness" via recognition).[21] In any case, the participant's identity is transformed through participation in the uprising.

The identity formed from belonging to a primary group is only one part of the insurgency. The individual motivations driving accession, integration, and operationalization are complex and unique for each primary group, from highly technocratic political organizations to loosely affiliated tribes. Tribal obligations are an interesting source of motivation and behavior in some irregular wars. Following the U.S. invasion of Iraq in 2003, a journalist interviewing Iraqi insurgents found that "many residents of the Sunni heartland believe there is no choice but to fight, provide shelter, or offer logistical help to the guerrillas who mastermind [anti-U.S.] operations—a reflection of the loyalty that is inherent in the tribal system operating in Iraq."[22] There are several compound motivations that are structurally viable in tribes, insurgent groups, militias, and so on regardless of the character of the group or the individual. These motivations include habitual obedience, ideological consent, opportunistic obedience, and contingent consent.[23]

Insurgents are often motivated to band together and conduct violent actions as a matter of perceived last resort. FLN leader Saad Yacef remarked during the Battle of Algiers in 1957 that "we are assassins. It's the only way in which we can express ourselves."[24] On the other side of the same conflict, the far-right French *Organisation Armée Secrète* (Secret Army Organization) expressed itself in similar terms: "Only violence will make us heard at present … it's the only solution remaining to us."[25] In an example from a different irregular war, Hamas activists employed suicide attacks and guerrilla raids rather than engaging in peace negotiations because its members felt that such talks were "useless and a waste of time" and "the enemy only understands the language of force."[26] The social power exchange mechanisms of consent and obedience can be divided along an axis of supportive attitudes versus supportive behaviors. Attitudes reflect motivation, while behaviors may be automatic, coerced, involuntary, and so on.[27] Figure 10 depicts these compound motivations combined with behavioral triggers present in insurgencies during irregular war.[28]

Insurgent motivations and behavioral triggers are constructed through the intersection of the types of commitment and compliance. Compliance derives from either consent or obedience, while commitment is shaped through ideology or opportunity. These combine in different

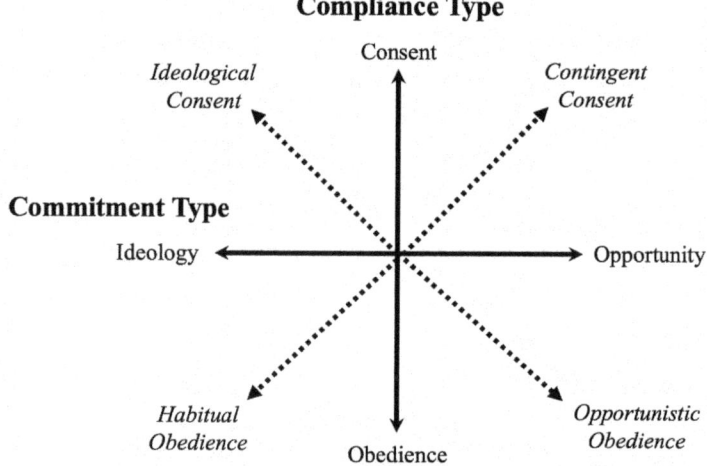

Figure 10. Insurgent Motivations and Behavioral Triggers.

ways to produce four motivational relationships insurgents often experience when participating in an insurgency. The least invested insurgent is opportunistically obedient: they participate for some immediate gain but cannot be relied upon to act on their own initiative for the cause. Next, the insurgent offering contingent consent is willing to participate because of some near-term issue they seek to resolve that is also sought by the insurgency. Moving into the more attached categories, ideological consent is a motivation derived from belief in the cause through the ideal prototype and ideal society the insurgent combines into an ideology. This motivation contributes to long-term group attachment as long as the insurgent continues viewing the shared ideological goals with the group. Even if the insurgent begins to grow disaffected through the divergence of the individual and group ideologies, the sunk cost the insurgent has already invested may exert an attractive force of its own upon the insurgent's decision to remain. Finally, habitual obedience occurs when the insurgent does not question the ideology of the group, instead being motivated by behaviors modified over time to produce instant, willing obedience to any order the insurgent believes the insurgency gives. These motivations and behavior systems of individual insurgents are bound up with the compound motivations that shape the primary group identity.

The Why

Often insurgent leaders and insurgents themselves clearly describe their compound motivations in memoirs, public statements, and other forms of communication. Souha Bechara, a member of the Lebanese Communist Party during Lebanon's civil war (1975–90) wrote in her memoirs that "there remains the basic cause for which I fought: a free Lebanon, a country at peace, but one also grounded in the ideals of justice and democracy."[29] Hamas founder Sheikh Ahmed Yassin spoke of his motivation for founding Hamas during his interrogation in Al Majdal Prison in Ashkelon on 6 June 1989. He noted that "I decided to establish a movement in Gaza to work against the Israeli settlement policy, resist the occupation and to encourage Palestinians to take part in the resistance effort against Israel."[30] Foday Sankoh's Revolutionary United Front (RUF) invaded Sierra Leone from Liberia in 1991. That group published a manifesto entitled *Footpaths to Democracy* clearly stating their *raison d'être* under the heading "What are we fighting for?" Their initial cause was clear: "We continue to fight because we are tired of being perpetual victims of state sponsored poverty and human degradation visited on us by years of autocratic rule and militarism."[31] Despite its idealistic beginnings, the RUF and its movement rapidly devolved into chaos and violence due in part to the political and economic institutions remaining intact as sovereignty crossed from the Sierra Leone government to the RUF.[32] Frantz Fanon's 1961 book *The Wretched of the Earth* inspired the RUF, as it did for other movements in Africa. The content and purpose of insurgent and anarchist communiques say more about the groups producing them than do the actions associated with them, such as damaging symbolic targets or cyberattacks.[33]

Those communiques or books may also serve as a sort of regional anchoring mechanism for unrelated conflicts in a region with shared deep history. For example, the 1971 Anya-Nya insurgency in Sudan made its platform clear in its manifesto, *Anya-Nya: What we fight for*, which listed the goals of representation, recognition, and self-determination.[34] The Anya-Nya insurgency was formed in Sudan in 1971, but it was a precursor to the Sudan People's Liberation Army (SPLA) and later variants that resulted in the formation of South Sudan from 2005 to 2011. The original goals in that 1971 manifesto remained remarkably consistent as generations of the insurgency progressed. Although SPLA leadership sought autonomy from Khartoum, most ordinary SPLA fighters were motivated by a desire for survival, both for themselves and the communities from which they came.[35]

In the case of Sudan, the differences in motivation from leaders and ordinary fighters were irrelevant as they converged upon similar goals. Parallels exist elsewhere as, for example, the interests of Hamas and the everyday Palestinian in Gaza coincide as both seek an independent Palestinian state and the return of the territory Israel gnawed away from them piece by piece from 1948 forward.[36] Similarly, Emiliano Zapata rallied his forces with the ideological phrase of *terra y libertad* (land and freedom) during the Mexican Revolution (1910–20).[37] The Cossack insurgency against Russia in 1708 made clear their purpose: "Our first task is to keep our land and hold it firmly."[38] Ukrainian Insurgent Army leader Oleh Martowych outlined the "Firm ideological foundations aimed at the political, spiritual, and social liberation of the people from foreign misrule"[39] for which his group fought. These declarations of motivation may not be universally accepted or applied within the group, but the pronouncements and personas from the leaders themselves certainly exert an outsized influence on group activity and character, while also shaping the ways states and other opposing forces attempt to engage with the group. Insurgent leaders often exploit relatively modest goals of the lesser members to mobilize forces toward achieving the broader objectives of the group.

Leaders in irregular wars often hail from humble beginnings themselves, despite the status they obtain in the primary group in the conduct of the insurgency. Many famously successful insurgent leaders of the mid-twentieth century were not military men but auto mechanics such as Ramon Magsaysay, librarians including Mao Zedong, attorneys such as Fidel Castro, and farmers like Abd al-Karim Qasim and Ben Ahmed Bella.[40] The Algerian revolutionary leader Ferhat Abbas was a pharmacist before the war,[41] and Envar Hoxha was a schoolmaster from Korce before he led the National Liberation Movement in Albania against the Axis and afterward ran the country for 40 years.[42] Che Guevara was a physician while Brazilian revolutionary Carlos Marighella was a civil engineer. Ahmed al-Awdeh, a former English teacher, led a major resistance force in southwest Syria after the Arab Spring. Sheikh Ahmed Yassin was a schoolteacher and a quadriplegic before founding Hamas in 1987.[43] Indeed, the other six founding members of Hamas were either teachers, doctors, or engineers. The archetype of an insurgent leader is not a politician or warlord but a representative, typified by the French-trained history teacher Vo Nguyen Giap.

The rise of Vo Nguyen Giap, one of the greatest military leaders in modern history, challenges the facile and specious notion of "the accidental guerrilla syndrome."[44] Instead, Giap's story serves as an example of the causal pathways leading to irregular wars. Giap joined the fledgling Communist Party in French Indochina in 1936, not as a militant but as a young Vietnamese student seeking to work within the existing political institutions to change France's oppressive policies in the colony. He authored a tract entitled *The Peasant Problem* in 1938, which was to become "the guidelines for subsequent communist policy in peasant regions" in southeast Asia during the wars against France, Japan, the U.S., and China.[45] Giap fled to China in 1940 after the Japanese assault southward toward the China-Indochina border. During this time, Giap's wife and infant child died in French prison and his sister-in-law was guillotined by the French in 1943. Giap's later decision to resist French rule became all the more personal after these deaths. While in China, Giap met Ho Chi Minh as the latter was teaching Kuomintang soldiers at a school in Kunming. Giap took command of the Armed Propaganda and Liberation Detachment in 1944, which consisted at first of only 34 men but then swelled to 10,000 in only six months. In as much time, Giap's newly formed Liberation Army had freed the seven northern provinces in Vietnam from Japanese control after partnering with the American Office of Strategic Services by request of Ho Chi Minh, a feat neighboring Thailand failed to accomplish even with significant unconventional warfare support from the Allies throughout the war.[46] After the Second World War ended and Japan left Indochina, Ho Chi Minh and Giap attempted to continue working with the U.S., but the U.S. chose to back France's doomed return to its lost colony. Thirty years later, Giap had defeated France, Japan, the U.S., and China. His group went on to achieve its original goal of independence, no matter the cost or the duration.[47] However, Giap's resolution of the "peasant problem" was not unique.

The cases of Hezbollah in Lebanon, the Kurds in Iran, and the Anya-Nya/SPLM in Sudan all illustrate the variations in leadership, group membership, and contexts of conflict in which these groups operated. At the same time, similarities in their characteristics become clear through comparison, such as motivations for primary group formation, leadership qualities, and movement objectives. These disparate cases demonstrate that insurgencies tend to develop amid sovereign dysfunction interacting with social order, sovereign territory, and political-economic institutions. They also represent the culmination of agitations expressed yet unreconciled among the civil society leading to attitudinal conditions supporting insurgency.

Sometimes social support to insurgency is rapidly exacerbated by external pressures, especially foreign invasion. Israel used *hasbara* (lit. explanation), the Hebrew equivalent to propaganda, to shape Israeli and American public opinion about Israel's reasons for invading Lebanon on 6 June 1982, temporarily concealing the Israel Defense Force's failed Operation Peace for Galilee.[48] Regardless of Israel's portrayal of the war through *hasbara*, the organized resistance force that would become Lebanese Hezbollah formed immediately in response to Israel's 1982 invasion of Lebanon.[49] However, Hezbollah decided to work within the state rather than against it following the signing of the Taif Agreement in 1989, which concluded the Lebanese civil war in 1990.[50] In 2008, Edward Said interviewed Hezbollah general secretary Hassan Nasrallah at the latter's home in Beirut. Following the interview, Said concluded that Hezbollah was still successfully resisting and defeating Israeli invasions of Lebanon, such as those in 1996 and 2006, because it remained a disciplined irregular force that melted into the population after combat operations rather than returning to military bases or carrying arms openly, mistakes the Palestine Liberation Organization paid for between 1970 and 1982.[51] In the 1980s, Hezbollah served two functions that its predecessor Amal could not: it positively resisted the Israeli occupation from 1983 to the Hezbollah-facilitated Israeli withdrawal in 2000, then chased Israel out of Lebanon again in 2006. Most importantly, Hezbollah provided socioeconomic relief badly needed among the Shi'a population in south Lebanon that had remained neglected by the state since the 1960s.[52]

Hezbollah capitalized on *muqawama* (persistent active resistance) to gain and maintain popular support with each subsequent Israeli attempt to weaken the organization, resulting in paradoxical gains for Hezbollah following Israeli operations against the group. This should be unsur-

Chapter 8. The Elements: People, Politics, and Propaganda   159

prising, as historical precedents exist in a range of contexts. Fidel Castro returned prisoners of the Batista regime without mistreatment, while Fulgencio Batista's troops routinely destroyed village homes, subsequently sending villagers into Castro's ranks.[53] Wendell Fertig, embedded with the Philippine resistance to Japan in the Second World War, commented that "in case of enemy occupation it is generally assumed that the civilians will resist,"[54] a fact Hezbollah understood in gathering support from non-combatants. Again, this support mechanism is not unlike Javanese resistance to the Dutch in the 1940s. Closer to home, one activist in Gaza who interviewed numerous witnesses to the fighting during the 2007 conflict with Israel noted that "the battle is really about the end of the occupation."[55] In Sudan, a witness to the war in Darfur opined that "if the government of Sudan wants truly to make peace, they have to provide security for the people. As long as they attack the villages or provoke others to do so, people will resist and join new groups. This is obvious to everyone."[56] Levels of insurgent resistance increase markedly when a foreign force occupies territory subject to the conflict.[57] Israel's July 2006 invasion of Lebanon underscored a simple, consistent reality: "occupation feeds resistance."[58] That 2006 episode illustrates well both the impetus for sympathetic support of Hezbollah and active resistance to Israeli forces leading to a transition among the civilian attitudinal conditions from uncommitted compliance with Hezbollah to sympathetic support.

Israel invaded Lebanon in July 2006 on a pretext planned months in advance based on a localized border skirmish between Hezbollah and the Israel Defense Forces. The Israeli invasion force consisted of about 30,000 soldiers versus a much smaller group of Hezbollah irregulars who held on for 34 days until Israel withdrew after killing 1,191 Lebanese civilians, injuring over 4,000, and displacing one million people along with deliberately bombing civilian infrastructure causing 4 billion dollars in damage mostly in densely populated urban Shi'a neighborhoods including the Dahieh suburb of south Beirut.[59] The Israeli Air Force flew over 5,000 strike missions in those 34 days,[60] while dropping over 4 million cluster bomblets throughout central and south Lebanon, of which 90 percent were dropped in the final 72 hours of the war after a ceasefire was already announced at the United Nations.[61]

Kristen Scheid was a professor at American University Beirut in 2006 and witnessed Israel's indiscriminate bombing campaign throughout July of that year. She reflected that "Israelis don't care if they are 'off-target.' They are even deliberately 'off-target' and hitting civilians at times to make us all surrender support for Hizballah. For me it has the opposite effect: I become more convinced that the only thing that will protect us is if Israel recognizes our parity, in might, in will to live, in ability to hurt our aggressors."⁶² During the Israeli invasion, IRGC-Quds Force commander Qassem Soleimani strategized the defense of the south Beirut suburb of Dahieh, along with Hezbollah leaders Hassan Nasrallah and Imad Mughniyah. Soleimani sometimes sat under trees, walked on foot, or planned operations in apartment buildings as Israel's faceless conventional offensive pounded civilian infrastructure and residents indiscriminately during their invasion.⁶³ In this context, it is not hard to understand how Hezbollah was able to not only gather new recruits for its militant wing, but also gain additional seats in Lebanon's parliament for its political wing following the 2006 Israeli invasion. In the 2009 Lebanese parliamentary elections, the first elections following the invasion, Hezbollah's political party won 14 seats in Lebanon's parliament, doubling its seats from the seven it held in 1996.⁶⁴ In south Lebanon, Hezbollah was able to frame itself as a functional sovereign in contrast to an external aggressor. Sociologist Jeff Goodwin put the point well in a broad sense:

> The continuous, indiscriminate repression of social sectors presumed to be sympathetic to the rebels will serve—however unintentionally—to prolong and perhaps even strengthen a mass-based insurgency, even if incumbents have introduced competitive elections and/or receive substantial foreign assistance. Indiscriminate state violence is especially likely to backfire, generating even greater levels of armed resistance, when states do not fully penetrate and control the territories they claim to rule.⁶⁵

Hezbollah is not unique in benefiting from the contrast in its own behavior toward civilians in war in comparison to the indiscriminate force of its conventional adversary. The Yazidis recognized this important distinction during their resistance to Ottoman and British incursions into their territory in northern Iraq and eastern Syria in the nineteenth and early twentieth centuries.⁶⁶ Their main enemy was the Ottoman government and its military officers, not the rank-and-file soldiers fighting for them. During one confrontation with the Ottoman forces in Sinjar at the end of the nineteenth century, one of the Yazidi resistance leaders admonished his men: "Keep hold of your Martini rifle, don't move the Mauser from your shoulder, do not fire on the rank-and-file, they are the children of the state. Look at all the ones whose sword hangs by their side, whose belt is sewn with gold and silver, throw those ones down."⁶⁷ The Yazidis sought autonomy and security, focusing carefully on their military opponent while casting their own plight in contrast to an oppressive external force, not unlike the Kurds in their attempt to secede from Iran in 1946.

The Kurds are the largest stateless ethnic group in the world.⁶⁸ The sovereign territory of Kurdistan transcends existing state boundaries, whereby northern Kurdistan (*Bakur*) is mostly in Turkey, western

Chapter 8. The Elements: People, Politics, and Propaganda   161

Kurdistan (*Rojava*) is mostly in Syria, eastern Kurdistan (*Rojhilatê*) is mostly in Iran, and southern Kurdistan (*Başûr*) is mostly in Iraq.[69] Kurds in Turkey, Syria, Iran, and Iraq have long sought autonomy, and the secession in Mahabad in 1946 was the closest they have come to achieving this goal in the modern period.[70] In the 1940s, the Kurds in Iran's northwestern province of Mahabad were a tiny community even within greater Kurdistan, having a population of only 16,000 in 1945, up from just 5,000 in 1890.[71] Regardless of the relatively small population, an insurgency in Mahabad simmered against the central governments in Tehran and Baghdad long before Mustafa Barzani led the Kurds to victory against two columns of Iraqi soldiers just over the border in 1932.

Barzani's forces began capitalizing on popular support and unified resistance to Tehran's decision to garrison 1,000 soldiers in Mahabad in the 1940s. The Shah deployed this garrison to maintain order between the Turks and Kurds, but the actual goal was to suppress Kurdish agitations for greater autonomy from the central government in Tehran.[72] Barzani used difficult terrain to confound the conventional doctrine of the British-trained Iraqis moving in from the west while also outmaneuvering the Iranian forces arriving from the southeast. Only after the British Royal Air Force carpet bombed civilian villages did the Kurds admit defeat in 1935, though they again defeated the Iraqis in 1943.[73] Their attempt at sovereignty did not end there.

On the political front, the Kurdish party Komala had been operating in secret with 100 members for eight months before going public in 1942, operating as "a clandestine organization composed of street cells."[74] Qazi Mohammad, an ally of Barzani, acted as the unelected voice and guide for Komala. Qazi Mohammad declared the independence of the Mahabad Republic of Kurdistan on 22 January 1946, and the first engagement with Iranian forces on 24 April 1946 was a Kurdish success.[75] Russian, British, and Iranian political power was absent from Mahabad in the mid–1940s, allowing Qazi Mohammad to wield power through his person in lieu of the state.[76] At the time of declaring independence in 1946, the Mahabad army consisted of only 70 officers, 40 non-commissioned officers, and 1,200 *sarbazi* conscripts, along with 1,200 gendarmerie rifles and ammunition from covert Russian shipments.

The Kurdish drive to declare independence in Mahabad was initially supported by the Soviet Union, who had already supported the independence movement in the Iranian province of Azerbaijan in 1945.[77] Russian advisor Colonel Kazimov oversaw training on drill, rifle marksmanship, machine guns, grenades, and the operation of 30 American and Russian vehicles.[78] Although Tehran crushed the insurgency by early 1947, the political alliance between Komala, Qazi Mohammad, and the Barzanis grew significantly in that period with the formation of the Democratic Party of Kurdistan in 1946, which later became the *de facto* and *de jure* power in northern Iraq with the Kurdistan Regional Government. An autonomy movement in Sudan achieved similar results.

The Southern Regional Government in Sudan was set up in 1972, structurally and functionally resembling the Kurdistan Regional Government in northern Iraq that was established two decades later, in 1992.[79] The SPLA insurgency was initially a state-parallel paramilitary that transitioned into its first conventional campaign in 1988 under Operation Bright Star.[80] The early success of that operation was due in part to the lack of centralized authority in the south, which led the government in Khartoum to rely principally on proxies for its war against the SPLA rather than fully committing the Sudanese Armed Forces. The SPLA reverted to irregular warfare after Omar al-Bashir came to power by coup in 1989. Omar al-Bashir changed Khartoum's strategy and employed the Sudanese Armed Forces in a large-scale conventional offensive in the south, relegating his proxy militias to a secondary role while actively sowing distrust within the SPLA. However, Bright Star succeeded in bringing two thirds of southern Sudan under SPLA control by 1990. Ultimately, the belligerents lacked control over the trajectory of the conflict in southern Sudan, in which the war of secession (1956–72) led to autonomy while the revolutionary war (1983–2005) resulted instead in secession.[81] In 2011, South Sudan was welcomed into the international community as the world's newest state.

Wings of Resistance

The structural elements of these movements share some commonalities tied to their functions, including political-coercive organs, administrative elements, and information management processes. The political-coercive organs consist often of a political wing, military branch, and sometimes a separate counterintelligence and security component. An administrative element usually manages finance, supply, logistics, and other activities required for the daily upkeep of the organization. An insurgency must also manage information whether for recruitment, disinformation against an adversary, internal messaging, or various other purposes. The responsibilities noted above can reside in any wing but the political and military wings almost always exist in an insurgency, no matter why the insurgency formed, what its goals are, or how it works towards those goals.

The political and military wings of an insurgency often operate with

Chapter 8. The Elements: People, Politics, and Propaganda   163

one wing subordinate to the other. The political wing most frequently holds the position of greatest influence. For example, the Taliban placed executive leadership in the *shura* council, its political body, while its military wing was subordinate to that authority.[82] Although political wings usually precede or dominate military wings in an insurgency, this is sometimes reversed. The Kosovo Liberation Army operated first as an irregular military force in Kosovo during and after Operation Allied Force in 1999, with the political component reaching prominence later.[83]

In Bolivia, Che Guevara oversaw another exception to the more frequent arrangement of the political element and non-combatant organizations relegated to an auxiliary role during the failed insurgency there. For the idealist Guevara, the main effort was guerrilla action in combat; all else was secondary. This exception, and the frequency of insurgencies who use it experiencing failure, indicates the possibility that the subordination of the military wing to the political wing may be the more effective arrangement.

The political wing is usually established first because insurgencies form out of failed political processes, when all political avenues of correcting sovereign dysfunction have been exhausted. For example, the *Partîya Karkerên Kurdistanê* (Kurdistan Workers Party/PKK) was initially founded as a political party in 1978, only transforming into a principally military movement in 1984 following the 1980 *coup d'état* in Turkey that shut the PKK out of Turkey's political system.[84] In El Salvador, the *Frente Farabundo Martí para la Liberación Nacional* (Farabundo Martí National Liberation Front/FMLN) formed in 1980 only after its constituents were subject to over 12,000 extrajudicial killings from far-right government forces. The FMLN formed after thousands had been killed for leftist leanings, and many of those who later joined were not initially directly affiliated with one another or with the political aims of the FMLN.[85]

The political wing in an irregular force often manages the primary group ideology, monopolizes intelligence, and spearheads popular mobilization all while attempting to penetrate the legitimate political system of an existing state or to form one of its own.[86] The political wing forms a security apparatus to employ repression and retribution while it also crafts messages to denounce opposition, gather support, and narrate the struggle in terms favorable to the movement to increase legitimacy. In other words, the political wing has three critical functions: retributive justice; eliminating opposition either by assimilation, expulsion, or purge; and constructing legitimacy using itself as the germ of a state.

The separation of an organization into discrete political and military wings in irregular wars is a common feature across otherwise disparate conflicts. The further designation of a military wing as a militia or a committee is another common element of irregular wars, with incumbent states tending to favor militias and insurgents tending to favor committees.[87] Regardless of the arrangement, nomenclature, or responsibilities these forces use, examples of political and military separation within insurgencies are a staple of insurgencies. In Vietnam, the Viet Minh formed the military wing while the National Liberation Front formed the political wing during the war against France.[88] The Greek resistance to Nazi Germany in the Second World War consisted of the National Liberation front as its political wing and the Greek People's Liberation Front as its military wing.[89] The resistance to apartheid in South Africa consisted of the Africa National Congress as a political wing and the *Umkhonto we Sizwe* as its military wing.[90] The Pakistan-backed Salafist insurgency in South Asia was broken into the military wing of *Lashkar-e Taiba* and the political wing of *Jamat ud-Dawah*.[91] The Irish nationalist resistance to British rule was split into a political party *Sinn Fein* and a military organization in the Provisional Irish Republican Army.[92] The autonomous Self-Administration of Northeast Syria, also known as Rojava, is a Kurdish government split into the political Democratic Union Party and the militarized People's Protection Units.[93] This pattern was also present during the Algerian War of Independence, in which the insurgency was split between the political *Front de libération nationale* (National Liberation Front/FLN) and the military *Armée de libération nationale* (National Liberation Army/ALN). This separation of responsibility is functionally logical and leads to bureaucratic efficiency, though specific organizational relationships, naming conventions, and delegations of responsibility may change for various reasons.

Political and military wings may also transform over time as conflict conditions pressure the organizational structure, as the example of Hezbollah demonstrates. During the earliest phase of its existence from 1982 to 1985, Hezbollah already had a military wing in the Islamic Resistance and a political body in the *Majles ash-Shura*. More mature political and military wings emerged in the form of an executive council and politburo with a subordinate military command after the Taif Accords in 1989 that ended the Lebanese civil war (1975–90). Despite these organizational changes, from its earliest days Hezbollah maintained regional offices conducting legal, fiscal, political, social, and military functions throughout Lebanon.[94] Since Israel's withdrawal from Lebanon in 2000, Hezbollah's political party has been the Loyalty to the Resistance Bloc while the Jihad Council oversees military activities.

Although the political-military structure is critical for the insurgency, there are other elements often strengthening the movement against the state and its security apparatus, notably the activists who are not actually

Chapter 8. The Elements: People, Politics, and Propaganda 165

a part of the primary group. Activists can be separated from the political and military wings of an insurgency.[95] Activists may run a press, fabricate necessary supplies, manage finances, run rat lines, facilitate communication, and provide other services as a support network. For example, although the political-military wings of the Ukrainian Insurgent Army were the "pioneers and directing force"[96] of the resistance to the Soviets and Nazi Germany, it was the activists who conducted most support activities outside the main organizational structure.

Successful organized violent resistance to a state or conventional forces in irregular wars often depends on the pivotal role of non-violent activities of the insurgency especially when proactively contrasted with the dysfunctional sovereign. Hezbollah in Lebanon in the 2000s and the Huks in the Philippines in the 1940s both had functional tax services, operated schools, published media, maintained hospitals, and so on.[97] There are four themes that are useful for understanding the role of militias like Hezbollah and the Huks in conflict resolution: the importance of context; understanding militia rationality and motivation; recognition and legitimacy; and organizational structure and strategy.[98] Sheikh Suwaydan observed that "Hezbollah is like communism," because its tenets include a central ideological doctrine, and a "long-term, futuristic scope," tied to the dual grievances of being "dispossessed" and deprived of rights by the state under which they live.[99] Likewise, the FLN/ALN in Algeria formed a state-like organization that conducted external affairs and internal security while holding committees for information, editorial activity, justice, finance, health, and trade unions.[100] All of this activity is designed to create a functional state, while no competitor can be tolerated even among allies.

A successful insurgency remains in a precarious position because it must continuously consolidate power, eliminate rivals, and establish credibility over its new constituents. All other parties must be eliminated, as happened in Iran from 1979 to 1989 whereby dozens of allied armed groups and political parties were purged, their members detained or executed, and their positions made impotent.[101] The Ba'ath Party followed a similar path from 1958 onward in both Syria and Iraq.[102] Islamic paramilitary organizations independently resisted Dutch rule following the Indonesian War of Independence, but these were quickly disarmed and then rolled into the *Tentara Nasional Indonesia* (Indonesian National Military/TNI).[103] Insurgencies craft messaging to justify these often violent power consolidations as a requirement based on the perceptions held among elements of society that a clear line can be drawn between supporters of the movement and its opponents.[104] However, the centrality of messaging is not limited to the final moments of an irregular war in which power is consolidated.

Ideology and the Message

Information is a powerful instrument in all contexts, and messaging in irregular war is no exception.[105] The importance of media to the incumbent in irregular war is revealed in the words of General Bello, who kept Peruvian dictator Alberto Fujimori in power by controlling access to information and suppressing an incipient social revolution: "If we do not control the television, we do not do anything."[106] Messaging is effective when it frames reduced costs with increased benefits in exchange for participation.[107] Messaging manifests in certain forms, such as intelligence information and propaganda. Visual media is especially powerful.[108] In 1953, the CIA employed Edward Bernays to run a propaganda campaign for the United Fruit Company to convince the American public of a fabricated threat of communist takeover in Guatemala. This provided justification for the 1954 CIA-backed *coup d'état* of the democratically elected Arbenz government that ushered in 40 years of irregular war so that "bananas and pineapples would continue to be safely picked by inexpensive native labor under careful watch, with all the profits flowing north," according to the U.S. government's chief propagandist at the time.[109] Propaganda contains the manifesto and the message, with further divisions based on the audience, the subject, and the medium.

The manifesto and the message form the ideological framework and justify the existence of the organization, while the message is designed for outreach, support, education, disinformation, de-escalation, and numerous other purposes. An insurgency in an irregular war often seeks to expand its base of support by conducting outreach activities, developing potential recruits, and advertising a positive image of the primary group while presenting a negative image of any opponents. Israel's *hasbara* (explanation) method of propaganda accomplished this task for the state, but irregulars operating in Palestine had already been putting this concept to use before the 1948 war. Former Israeli Prime Minister Yitzhak Shamir described the importance of *hasbara* for raising operating funds for his Nazi-aligned *Lehi* (Stern Gang) which he led in Palestine during its insurgency against the British in the 1940s: "We badly needed financial resources to build up bases, to buy arms, to pay for our broadcasts and pamphlets, to keep ourselves alive."[110] During the Battle of Algiers in 1957, FLN leader Ramadane Abane opined that "our struggle must become known. We could kill hundreds of colonialist soldiers without this ever being announced."[111] The conflict has little hope for survival without effective propagation of the group ideology to new, receptive audiences.

Some disagree on the exact meaning of *ideology,* perhaps because not all people respond uniformly to evaluative information from the same

Chapter 8. The Elements: People, Politics, and Propaganda   167

origin. For example, political ideology could be "a set of values or attitudes oriented about the problems of the state"[112] or ideology in general may be just "a normative theory of action."[113] Regardless of the exact definition, ideology is central to the process of information transmission in irregular war. Strongly ideological individuals begin with an ideology or schema to form an attitude, while less ideological individuals favor simpler symbolic predispositions to form attitudes.[114] Attitudes are shaped through expectations, symbolic inferences, and worldview.[115] Attitudes, values, and beliefs work together to shape behavior preferences in individuals as thinkers and actors. Importantly, these attitudes are all acquired through learning processes, implying they can be reversed, modified, or imposed. Attitudes impact behavior, while intention strongly predicts performance.[116] Attitudes, behaviors, and intentions all affect individual ideologies.

Ideology is spread easily to receptive audiences, and perhaps one of the most conducive environments can be found in prisons.[117] During the Iranian revolution, Ayatollah Khomeini expressed that "prison must become like a university for training humans."[118] Similarly, the French war in Algeria ensured that prisons would become "a marvelous recruiting and training center,"[119] each cell becoming an informal academy for spreading ideologies against the state. In fact, one Algerian FLN minister noted after the Algerian war of independence that "more than 30 of us were arrested during the electoral campaign [of 1948 in Algeria before the war] and put into jail for years. A look at the list of those jailed gives you an approximate list of the actual leadership of the Algerian revolution."[120] A similar situation arose out of the U.S.-run Camp Bucca in Iraq, from which sprung numerous leaders of what would become Da'esh after the U.S. withdrew in 2011. Those who have behaved counter to the laws, norms, and traditions of the state, while present in a correctional institute, are even further susceptible to radicalization than if they had not been incarcerated.[121]

Accidental detentions and detainee abuse by conventional forces during a counterinsurgency operation can also easily turn those who were indifferent or beholden to that force into enemies. For example, American forces in Iraq detained Haider Saber al-Abadi for six months in Abu Ghraib, subjecting him to sexual assault and physical abuse despite being an opponent of the regime the U.S. had just overthrown, much like the "70 to 90 percent of the persons deprived of their liberty" who "had been arrested by mistake" during the first two years of the American invasion and subsequent occupation after 2003.[122] Al-Abadi described his experience during an interview: "Whatever happens to these soldiers, it will not give me back my dignity. Who will return this respect to me? My honor has been crushed underfoot."[123] Ideology may also be spread somewhat passively, especially through memoirs and manuals from previous conflicts.

Individual memoirs or manuals from irregular wars serve as historical documents discussing the procedures of resistance, the precepts of rebellion, and the philosophies of revolution: these declare the ideology of the particular actor concerned. Their tenets often find audiences far beyond their original context.[124] Insurgents find inspiration in the writings of those who fought for alien ideologies and in different contexts of conflict than their own, while tactical and operational literature exists for different sides of the ideologies separating the belligerents. For the Spanish Civil War (1936–39), Peter Kemp wrote from the Nationalist perspective while George Orwell and Ernest Hemingway wrote from the Republican side.[125] For urban operations in irregular warfare, Swiss soldier von Dach's manual on right-wing insurgent resistance to a Communist invasion of Switzerland is not dissimilar to Brazilian Communist insurgent leader Carlos Marighella's own manual for urban resistance to a right-wing regime.[126]

Operations and tactics are generally independent of ideology. Manuals written by Marxists such as Che Guevara in Cuba or Carlos Marighella in Brazil are useful to an Islamist *jihadi* fighting American forces in Iraq 50 years later. Sudanese SPLA/M leader Garang was inspired by Marxist-Leninist military doctrine. Garang learned of this doctrine during the military support his group received from the Derg in Ethiopia until 1991. The imprints of Marxist-Leninist ideas of war were clear in Garang's decision to move from guerrilla war to conventional war, and then back again. Similarly, the unconventional warfare guides by the Nazi officers Ehrhardt and von der Heydte were taken with the Nazi High Command's 1944 *Bandenbekämpfung* (Combating Bandits) manual to form some elements of the doctrine used in the special operations forces of the U.S. and United Kingdom.[127] Meanwhile, Omar Cabezas published *Fire from the Mountain,* telling his side of the story of the Sandinista-led revolution in Nicaragua (1961–79), a book which provided the only insight most foreigners had regarding the conflict in the subsequent decade.[128] The Provisional Irish Republican Army published its *Handbook for Irish Volunteers* in 1956 as an ideologically neutral tactical guide to urban insurgency, while their propaganda themes of repression, resistance, armed struggle, and physical action remain applicable in unrelated conflicts.[129] Al-Qaeda founder Abdullah Azzam wrote *Join the Caravan* as an outreach device to generate support for the war against the Soviets in Afghanistan (1979–89),[130] while al-Qaeda also adopted the writings of Clausewitz, Mao Zedong, and Taber almost without alteration into their own doctrine.[131] Memoirs and manuals do much to generate exploitable sentiments within the primary group in an irregular war, while leaflets serve a different purpose.

Leaflets are a form of propaganda designed to reach audiences outside

Chapter 8. The Elements: People, Politics, and Propaganda 169

the primary group but within the same context of conflict, including potential supporters and adversary military elements. These may galvanize support from the people, seek to alienate them from the adversary, or demoralize adversary forces. The importance of reaching those within the context of conflict is illustrated in the four fronts of the Javanese war against the Dutch: guerrilla war; war for popular support, specifically civilians uncooperative with the Dutch; propaganda and diplomacy; and the war for support of Indonesians and others abroad.[132] Note that three of the four fronts are related to information and propaganda, while only one involves actual combat. Leaflets were used extensively in many irregular conflicts over the last four centuries.

The common use of leaflets and other printed media actually predates the modern sense of the word *propaganda*. The current use of the term *propaganda* arose when the Catholic church established the Office of the Propagation of Faith in 1622.[133] That first official propaganda body served a similar purpose to Islamic *da'wah* (proselytizing) activities. Islamic *da'wah* plays a crucial role in recruitment, sustainment, and support for some modern conflicts, including those in Iraq, Syria, Indonesia, North Africa, and elsewhere.

Leaflets and other printed propaganda long played an effectively secular, rational role from the earliest days of their use. In 1606, Bolotnikov's Cossack insurgency smuggled thousands of leaflets into Moscow, calling for the poor to rise against the wealthy, pledging important offices after the insurgency succeeded, and promising the insurgents could help themselves to the wives and wealth of the conquered.[134] Leaflets distributed during Bulavin's insurgency in Russia in 1708 described the methods the Cossacks sought to use to achieve their goals: "Annihilate the boyars, Germans, and profiteers … plow for yourselves."[135] Leaflets have changed little in their form and messaging style even as print media has given way to digital and social media. Effective use of propaganda means the difference between an insurgent and a freedom fighter, between an invasion and a humanitarian intervention.

One purpose of propaganda is to establish solidarity and accumulate distant sympathizers to nurture, maintain, and strengthen the cause while justifying resistance regardless of the medium.[136] Joseph Stalin considered publication of propaganda through pamphlets, newspapers, and other printed media essential "to create a solid core of the party capable of uniting the innumerable circles and organizations into one whole, to prepare the conditions for ideological and tactical unity, and thus to build the foundations for formation of a real party."[137] Mao Zedong argued that "propaganda materials are very important. Every large guerrilla unit should have a mimeograph and a printing press."[138] Fidel Castro used propaganda and the exploitation of fear to convince the Batista government in Cuba that an invasion was imminent, when the irregular force amounted to only several hundred men with just one .50 caliber gun as their heaviest weapon. In response to this successful deception through propaganda, Batista mustered over 5,000 troops in response to conduct static security along with cordon and search missions in a jungle so dense they could not communicate with adjacent units.[139] In the same conflict, Che Guevara used propaganda to explain what he framed as "the theoretical significance of the insurrection"[140] in the form of a serial publication. It is this latter method that deserves special mention. Indeed, the serial publication is perhaps the most ubiquitous propaganda medium found across irregular wars.

Insurgencies use serial publications to promote their cause, highlight successful operations, denounce adversaries, and recruit supporters. In the 1940s, the Ukraine Insurgent Army's underground magazine was *Povstanets* (The Insurgent),[141] while the Paris-based anarchists printed *L'Anarchie* (Anarchy) to communicate to readers that the movement required "a turn from individualism to social action."[142] Starting in 1971, the Mosaddeqist National Front based in Lebanon wrote *Iran al-Thawra* (Revolutionary Iran), which provided Farsi translations of Latin American Marxist-Leninist writings such as those from the Tupamaros (MLN-T), Chilean *Movimiento de Izquierda Revolucionaria* (Revolutionary Left Movement/MIR), and Che Guevara.[143] Bismarck banned the Social-Democrat Party in Germany in 1878, but this did not stop the group from clandestinely publishing the magazine *Der Sozial-Demokrat* (The Social Democrat).[144] The Provisional IRA published its clandestine newspaper in Northern Ireland called *An Phoblat* (Republican News).[145] A printing press arrived in northwestern Iran in November 1945 as a gift from the Soviet Union to Mahabad, and the paper *Kurdistan* began immediate circulation.[146] Mirza Kuchik Khan's *Jangali* (Jungle) irregulars in Gilan, Iran printed their newspaper entitled *Jangal* starting in 1915.[147] The media organ for the Nazi-aligned Jewish *Lehi* (Stern Gang) in Palestine in the 1940s was the *Front de Combat Hebreau* (Hebrew Combat Front), while Frantz Fanon edited the FLN newspaper *El Mudjahid* (The Struggler) in Algeria beginning in June 1956.[148] The Office of the Martyr Sadr, precursor to the more robust *Jaish al-Mahdi* (Mahdi Army) in Iraq, published the newspaper *Al-Hawza* (The Shi'a Seminary) until the U.S. military shut their press down in March 2004.[149] Similarly, the Services Office, a precursor to al-Qaeda, published *Jihad* (Holy War) after 1985. As can be seen, these publications often carry in their name an overt nod to the group's *raison d'être*.

Not all serial publications are meant for the public, however. Some

Chapter 8. The Elements: People, Politics, and Propaganda 171

are published under misleading titles to avoid scrutiny or to deceive the incumbent security apparatus. For example, the Red Army Faction/Baader-Meinhof Group published the innocuously titled *Red Book 29* in West Berlin, while the French *Maquis* manuals from the Second World War circulated as the *New Traffic Regulations* and *Instructions for First Aid*.[150] Similarly, the delivery method may change to reach a specific type of audience or increase reach. The digital publications such as Hezbollah's *Manar* (Lighthouse), Da'esh's *Dabiq* and *Rumiyah* (Rome), and al-Qaeda in the Arabian Peninsula's *Inspire* may be electronic but their content and function remain unchanged from the print media format of other organizations.[151] Ultimately, the message and the manifesto are but two arms of a group's ideology.

A complete ideology contains a critique of a competing idea, affirmation of accepted concepts, and strategic guidance.[152] That ideology is designed to expose the perceived inadequacy of an existing order, describe a better alternative, and then explain how to get from the existing order to the new one using the primary group as the only conduit to reaching that goal. A slightly different set of themes for revolutionary ideology begins with the message that opponents are incompetent, impotent, and evil. Second, victory is inevitable, while prudent optimism should prevail. Finally, the revolution is shrouded in universal truths that transcend the crisis.[153] The elements of a successful ideology include straightforward objectives, a rallying component often expressed through a charismatic leader, a clear enemy (e.g., a person, idea, or system), a requirement to recognize a sole legitimate ideology, and an organizational philosophy.[154]

There are two key lines of inquiry one may take to understand an ideology. The first concerns the ideal society and the second concerns the ideal person according to that ideology.[155] The importance of ideology to organizations such as Hezbollah is clear. In Hezbollah's case, ideology is passed through formal mechanisms such as the *hawzah* (Shi'a seminary) system of learning centers operating in Lebanon. Hezbollah does not run all *hawzah*, founding only three of the 15 *hawzah* in Lebanon.[156] Of those that Hezbollah does control, each *hawzah* is led by one *sheikh* (religious scholar) overseeing 40 to 50 students.[157] Hezbollah uses a three-phase process to assimilate the outsider into the primary group. First, *da'wah* (proselytizing/outreach) gains recruits who are then indoctrinated in the *hawzah*. Through the *hawzah*, the recruits start on the path to becoming a *hezbollahi* (party member). The ultimate objective of this process is to create cohesion through *'asabiyyah* (group-feeling). A *marjah* (one to be emulated) provides orientation through their ideal prototype for the *hezbollahi* to follow throughout their tenure in Hezbollah.[158]

The development and dissemination of ideology through the people, politics, and propaganda in irregular wars is distinct from the conventional forces and states involved in the same conflicts. Primary groups are defined by civilian attitudinal conditions toward the insurgency, individual *'asabiyyah*, and the motivations and behavioral triggers pushing a movement from passive to active resistance. Memoirs and manuals from specific conflicts and moments in time carry a different message than the doctrinal publications, traditions, and laws governing conventional forces. It is useful then to explore how conventional forces and insurgents intersect in irregular wars.

CHAPTER 9

A Dialectic of Irregular War

> It is not until there is resistance that there is war.
> —CARL VON CLAUSEWITZ[1]
>
> No states, no revolutions.—JEFF GOODWIN[2]

Why do armed citizens decide to engage in organized violence against a state when the costs are irrationally high and the outcomes impossibly uncertain? The central issue in Abdolhossein Zarrinkoub's *Two Centuries of Silence* revolved around an anacoluthon in Persian history in which weak, poorly equipped, and loosely organized bands emerged from the desert in the eighth century to overthrow a great power with two thousand years of history. Importantly, the Muslim conquerors framed their conflict in Manichaean terms, pitting the *mushrikun* (non-believers) against the new religion and its *ummah* (global community of Muslims). Zarrinkoub asked, "How was a world civilization with so many achievements in art, architecture, religion, law, literature, et cetera destroyed by a nomadic people with limited literacy and few accomplishments?"[3] A similar non sequitur occurred in the Westphalian context along the road to the Battle of Borodino in 1812 and another in the midst of the Peninsular War (1807–14), in which the greatest military leader Europe had known since Caesar lost in the first to pride and partisans and in the second to perfidious Albion and *guerrilleros*. In his 1886 poem *Arithmetic on the Frontier*, Rudyard Kipling searched his experience in Afghanistan for a formula to solve the problem Zarrinkoub raised, a conundrum in which "two thousand pounds of education drops to a ten-rupee jezail."[4]

The Intersection of People and the State

This book has framed such questions in their convergence between a dysfunctional sovereign and social, territorial, and political-economic conditions. However, the existence of those conditions is not enough for insurgency to develop; elements such as people, states, and third parties also interact under these conditions, and this complex set of interactions may lead to irregular war. Put simply, irregular war is a synthesis of political violence between people and states, in which the people surge upward against an incumbent sovereign to change, displace, or transform structures of power. Explaining the onset, conduct, and resolution of insurgencies requires first understanding the identities, motivations, and means available to those comprising them. This requires asking who they are, what they want, and what they are willing to do to get it.

Insurgencies and general revolutionary movements tend to operate within Westphalian dictates, opposed to a *particular* state but not to the *idea* of the state. Indeed, ideologies profoundly affecting the world order such as communism, religious extremism, nationalism, self-determination, and secession have shared an affinity for joining rather than overturning that international system of sovereignty and mutual recognition. Despite the vitriol of certain ideological tracts to the contrary, the insurgent's ultimate goal is most often recognition in a Hegelian sense. This means recognition via social justice, state sovereignty, and inclusion in political-economic institutions, a process which political scientist Francis Fukuyama framed with the Platonic concepts of *thymos* and *isothymia*.[5] These two concepts refer to something Fukuyama inferred from Hegel and Plato, both in terms of something we can call "spiritedness." This *thymos* or spiritedness is linked closely to the individual desire for recognition from a group: it is the give and take of social proof.[6]

The question of who the insurgents are and where revolution should begin is contentious, even among the participants. Vladimir Lenin and Joseph Stalin favored the proletariat, but Mao the peasantry. Ali Shariati looked to the intelligentsia while Ayatollah Khomeini idealized the clergy. The Mensheviks sought to preserve an element of the old privileged middle class, while Peter Kropotkin, himself an aristocrat, sought to destroy it. Argentine Marxist Che Guevara and Jordanian Salafist Abu Musab al-Zarqawi both sought to generate support for their movement through an artificially instigated violent momentum, the former in the Bolivian countryside and the latter in the urban jungles of Iraq. These differences in ways and ends are dependent upon means, just as in any other political-military conflict.

Some argue that violent political factions, armed groups, and insurgencies are an expression of a disrupted patron-client system of rents exchanged for protected privileges.[7] It is not coincidental that most belligerents in irregular wars begin first as non-violent political parties before resorting to force. This is because, as French West Indian anti-colonialist Frantz Fanon observed, "Political parties never insist upon confrontation precisely because their aim is not the radical overthrow of the system."[8] The costs of transitioning from a non-violent political process to the violent dialectic of irregular war are immense and there is a low probability of success at the outset.

There must then exist a perception among the potential insurgents that such a transition is right, possible, and the only option. Goal attainment strategies typically seek outcomes with the highest reward and least effort.[9] Insurgents calibrate this motivation when seeking change through the only means perceived to be effective, including violence. This involves stepping outside the existing institutions of the state to achieve a social, economic, or political goal, often requiring force to do so. It is only when those social institutions fail consistently, egregiously, and blatantly that citizens *opt out* of the political process and *opt in* to political violence.

Participants in such violence form an identity galvanized in opposition to an incumbent authority. This can occur through the catalyzing effects of suppression, reaction, and exclusion from the incumbent sovereign, transforming individual feelings of alienation, dissatisfaction, and marginalization into active resistance expressed through an opposition group. Violence is then only a means; justice as fairness is the true end. Perceptions of what is just and what is fair are subjectively shaped in each social context.

Many factors uniquely influence social contexts and define what citizens expect from the functional sovereign. These factors include forms of government, religions, languages, financial systems, agricultural practices, physical geography, political institutions, or any other structures creating, maintaining, or destroying a social equilibrium. These factors influence how citizens perceive the state's responsibility to provide expected outcomes. If those expectations are unmet, people seek political change to bring outcomes in line with expectations.[10] Insurgencies may emerge in an environment in which votes cannot produce the political changes demanded.

An insurgency is a political party expressing its vote through the violent rejection of the incumbent when that choice is perceived to be unavailable in the current structure but the desire for change nonetheless persists. North Vietnamese defector Le Xuan Chuyen defined the party as "a doctrine, a theory, a belief."[11] If a critical mass of relevant people believe that the political process will address their grievances, they will be less likely to seek redress for those grievances outside the political process in existing institutions versus through protest, resistance, or other violence. If a party reflects as a prism those functions of doctrine, theory, and belief while also redressing group grievance, it is unsurprising then that insurgency develops often from political parties whose constituents perceive relative exclusion from state processes. In all cases of collective violence, non-violent activity usually precedes the violence.[12]

To qualify collective violence as war, the violence must be highly coordinated and the damage highly salient relative to other forms of contentious politics.[13] The introduction of violence specialists and political entrepreneurs such as police, militaries, and so on are drivers of rising levels of coordinated destruction.[14] The terms "political entrepreneurs" and "violence specialists" used in political science feature a parallel construction to Shariati's thinkers and actors, with these entrepreneurs and specialists steering the direction of the coordinated destruction while not always remaining in control of it.[15]

Three major forms of coordinated destruction include campaigns of annihilation, conspiratorial terror, and lethal contests.[16] Displays of coordinated collective violence are groupings of performances that replicate recognition in some way, especially when these performances are violent rituals.[17] While violent rituals and coordinated destruction certainly feature in irregular wars, these are best used as descriptive terms for specific violent episodes rather than as explanatory gates. Instead, it should be clear by now that the ultimate objectives of insurgencies in irregular wars are specifically political rather than military.

The objective of the insurgency is fundamental political change outside established institutions. Insurgents seek to synthesize political legitimacy and popular will into a new social order, sovereign territory, or set of institutions. This synthesis is necessarily violent, as all non-violent means are presumed to have been exhausted prior to the onset of an irregular war. The breakdown of a dysfunctional sovereign beyond the capability of providing collective protection usually leads to the spontaneous formation of grassroots self-defense units.[18] This devolution of responsibility for provision of public goods from the sovereign to the local community is an important factor in the transition from non-violent resistance to active, organized violence against the state.

A resistance movement such as an insurgency is composed of three parts. First, shared grievances bind the group and attract new members. Second, the group defines itself by its opposition to an enemy, which is often either a state or an instrument thereof. Third, the group has a shared political-military vision, often expressed through the thoughts and actions of its leaders. For example, elements at work in the 1979 revolution in Iran could be divided into the role opposition leaders played, the nature of opposition organizations, the potential for mass mobilization, and the incumbent regime response to the first three elements.[19] Collective action in a primary group is constrained by environmental conditions, tradition, time, institutions, education, and other factors.

Chapter 9. A Dialectic of Irregular War 177

The perceptions of justice and fairness held within the group may lead a group member to believe their actions are completely justified despite the objectively criminal or taboo nature of those actions within the incumbent legal or social system.[20] These beliefs are a product of the imagery, illusions, experiences, and lines of reason expressed within the primary group against the incumbent sovereign. The beliefs are strengthened through processes such as affirmation,[21] repetition,[22] and stickiness.[23] Participants in a resistance movement are bound through beliefs, collective actions, and shared responsibilities all toward the objectives they perceive to be possible only through organized violence against the state.

Structural differences between insurgencies in various irregular war cases do not detract from the similarities among these groups, especially in their objectives. The South Sudan White Armies were often nothing more than small bands of armed youths defending their villages and livestock from the state, while the Ukraine Insurgent Army carried arms openly, wore uniforms, and organized under a clear chain of command under a general staff headquarters.[24] Despite these comparative differences, one South Sudan White Army militant revealed his analogous purposes in remarking that "the White Army only existed because there was no government."[25] Similar conditions existed in Malaya in 1946, albeit in a different context, in which the immediate breakdown of a functioning sovereign lay the foundation for the Malayan Emergency (1948–60). In such cases, the objectives were political, focusing on establishing a functional sovereign to stabilize the social order, protect territory, and provide inclusive institutions.

Chief among these objectives is the just distribution of political rights and the fair access to economic processes. Unjust political systems combined with unfair economic processes often lead to conditions of relative poverty. This poverty becomes a critical factor when the general population begins to demand that the "narrow elite" stop structuring society in ways that only benefit the few.[26] Some irregular wars in the twentieth century resulted from this general poverty combined with exclusion from political-economic institutions, as traction grew over grievances about unequal access and compensation for fuel, labor, and materials in extractive economies.[27] Insurgencies overwhelmingly seek to overthrow the incumbent regime while keeping the bounds of the state relatively intact, as opposed to the comparatively less frequent attempts at secession or combination into new states.

Some theorists and military practitioners have failed to link these political-economic conditions to the onset of irregular wars, instead allowing tactical myopia to distort the bigger picture. These views usually propose that irregular wars occur in difficult terrain such as jungles, deserts, and mountains, areas sharing the common theme of sparse population and distance from urban centers under state control. This was a popular view among those writing during the Cold War, in which most military adversaries were Marxist-Leninist guerrillas. However, relegating those adversaries to isolated, difficult terrain is a narrow description of one type of tactical employment of irregular forces conducting irregular warfare rather than irregular war as a whole. Some understood this synecdoche trap even during the Cold War, as the leader of the Tupamaro guerrillas in Uruguay advocated the urban centers as his main focus for warfare, not unlike the Irish resistance to Britain or the Ukrainian Insurgent Army. The Tupamaros used Carlos Marighella's *Minimanual* as a doctrinal publication for their movement from 1962 to 1972.[28] Marighella's emphasis on urban rather than rural insurgency fit the Tupamaro *modus operandi*, and therefore appealed more than other ideas available to them, such as Guevara's *foco* theory of rural-to-urban warfare.

Popular struggles, revolutions, and resistance movements are forms of mobilization of one ideology against another, as insurgencies seek a utopian outcome for the *novo homo* promised in their distinct ideology.[29] The incumbent fears what appears to be a zero-sum game between the sovereign and the ideal, and so states opt to suppress a violent challenge from below with violence from above. It is thus that states too often attempt to solve social problems, however violent, with a military instrument.

The State and Its Enemies

The form of complex society that has become the modern state is essential for irregular war, and the modern state is a unique addition to the story of human societies.[30] Despite the prevalence of both Westphalian sovereignty and irregular wars, states have only been a major political factor in human society for one-fifth of one percent of human existence. Proto-states first emerged around 3300 BCE in Mesopotamia, coalescing in their current form around 1650 CE, a boundary that is relatively stable for marking the beginning of the dominance of the state in society worldwide.[31] The sovereign state is a brand new way of organizing human societies in the long history of our species. The existence of the Westphalian state is a necessary condition for the kind of insurgencies and revolutions that have occurred globally since the seventeenth century. This relationship between the state and insurgency is affected by the expansion and dominance of the international political system.[32] This system combines with state-level factors contributing to revolutionary movements and violent insurgencies, with likely candidates being states that are patrimonial, weak, and repressive. Weak states are characterized by relatively anarchic

Chapter 9. A Dialectic of Irregular War

social organizations existing outside the control of one or a few extractive urban cores, such as exists in Niger, Mexico, Sudan, Syria, Pakistan, and other places.[33] Whether the state is strong, weak, or anything in between, irregular war does not exist without the conditions the sovereign state produces in relation to the people under its shadow.[34]

A specific meaning is intended in the use of the term *Westphalian state* that applies when analyzing irregular wars over the past four centuries. One definition holds that the state is "a people living under laws," while a looser definition holds that the state is an administered polity that can be modified by a recognized authority.[35] A more nuanced definition holds that the state is "An autonomous political unit, encompassing many communities within its territory and having centralized government with the power to collect taxes, draft men for work or war, and decree and enforce laws."[36] Some have argued that Max Weber's condition of the state exercising a monopoly on the means of legitimate force was an insufficient criterion on its own to define the state while at the same time acknowledging that this force is a necessary component of Westphalian sovereignty.[37] In synthesizing these ways of understanding Westphalian states, we can see that social order, national territory, and political-economic institutions are all elements affected by dysfunction within the system of sovereign coercive power over the people.

The importance of the ubiquity and uniformity of Westphalian state sovereignty at the level of relations between states cannot be overstated. Such a form of sovereignty provides the limits of social orders, boundaries of territory, and structures of political-economic institutions essential to the state and its participation in the international forum of nations. This concept of the state was, in effect, internationally ratified during the 1933 Montevideo Convention and formalized through the Charter of the United Nations in 1945. Indeed, borders are not so valuable in their intrinsic ability to physically separate states, but instead because they are a physical artefact of agreements between states, acting as beacons of sovereignty and reminders of autonomy to other states.[38]

It is no surprise that states forming from sovereign structures alien to European political philosophy eventually adopted the Westphalian concept of state sovereignty with gusto. While the cases of China, Pakistan, or Indonesia may come to mind first as illustrative cases, two others deserve mention. Two other former imperial powers, Iran and Turkey, viewed the Westphalian system and attendant internal political dynamics as the only option following the Axis loss in November 1918. Specifically, post–Axis Iran and Turkey both understood that entering the club of modernity required consolidation of power in a central government based on the industrialized French state following the Napoleonic wars.[39]

This emphasis on a Napoleonic authority structure and a centralized state on the European model has taken the world by storm. Fully 97 percent of the world's current international borders were established after 1700, while over 50 percent were established only in the last century.[40] Access to open water is also apparently a major factor affecting conflict and state formation. For example, of the roughly 195 sovereign states, 44 are landlocked, and 9 of those 44 have the lowest human development index in the world.[41] A doubly landlocked state is even rarer, with Liechtenstein and Uzbekistan being the only examples. Ungoverned, autonomous, and contested areas still exist, but they are the exception and are often violently disputed or practically inaccessible.[42] The current system is a logical outcome of the concentration of populations in discrete territorial arrangements under centralized authorities, since dispersed and separately governed populations are harder to defend, share less in common, and have reduced incentives to cooperate.

Sovereignty takes on a specific meaning in irregular war. State sovereignty is a combination of the government's power, force, and authority expressed through executive action, law, and justice, respectively.[43] In the ideal Westphalian state, power and force reside in the law while authority resides in justice. State power is either latent or military in its expression, with a preference for either depending less on logic and more on the actual capabilities the state can muster in terms of men, weapons, and equipment.[44] Sovereignty is an attribute of a ruling entity, and state sovereignty is a combination of national instruments ideally wielded in the public interest.[45] Given these elements of sovereignty and the state, it makes sense to describe war as an activity that contests power, legitimacy, and justice. The essential elements of a functional justice system are courts, prisons, and police.[46] The contest over those functions expressed between states is conventional war, while irregular war is a contest of power, legitimacy, and justice involving a belligerent that is not a state.[47] These contests of power and authority are more likely in some states than in others.

The differences are especially clear when categorizing states according to the relationship between the people and the government in a particular state. Writing in 1920, French jurist Raymond Carré de Malberg devised a useful framework to categorize these states. His framework included three types of state: *état légal* (legal state), *état de police* (police state), and *Rechtsstaat* (constitutional state).[48] The *Rechtsstaat* contrasts with *l'état légal* and *l'état de police*, as shown in Immanuel Kant's "doctrine of the state based upon law ... or of 'peace through law,'"[49] with rule of law being a prerequisite for economic development and market security. Further, political economy functioning in ideally institutionalized states generates information, allows people to believe in the state's claims for their wellbeing, encourages improvement through policy, and contributes to shared knowledge about the state's intentions.[50] In this sense, political economy is the interaction between pursuits of wealth and power, pursuits available to all in the *Rechtsstaat*.[51] Irregular war is unlikely in the *Rechtsstaat*, since it features pluralistic societies, rule of law, and inclusive institutions.

States with pluralistic societies protected by rule of law are least susceptible to civil war, social revolution, and political violence.[52] Rule of law ensures equality before the law in legal matters, constrains the elite, and protects a majority otherwise vulnerable to a powerful minority and vice versa. The *Rechtsstaat* provides an institutionalized process of managing grievances through inclusion rather than coercion, and rule of law is a prerequisite for economic development and market security. Economic growth alone is not necessarily a strong correlate for reducing instability since an oppressive ruling elite in an *état légal* can become enriched with private benefits legally acquired as actual revenues increase despite general poverty prevailing among the majority. In this case, people are persuaded rather than coerced to work with and for the state. Persuasion is a voluntaristic and intentional form of social influence that a functional sovereign uses to induce citizens to comply with and defend the state.[53]

Irregular war is also less likely in voluntaristic states compared to coercive states. The idea of broad voluntaristic and coercive categories of states holds that voluntaristic theories of the state are less compelling than coercive ones.[54] Prominent voluntaristic theories were debunked soon after they were proposed, such as Karl Wittfogel's hydraulic theory of voluntaristic state formation.[55] This may be due to an individual choosing to affiliate with the state and then participating in a formal process of redressing grievances, diffusing much of the conditions leading to organized violence against a state. Voluntarism does not always equate to liberal democracy, and liberal democracies experience varying degrees of stability depending on their development and other factors. Democracies in newly established states that have not yet consolidated face the threat of *coup d'état*, collapse, or reversion to dictatorship if the redistributive effects of democracy grow too great to bear, especially for the elites of the *ancien régime*.[56] Although a *coup d'état* is not equivalent to irregular war, revolutions from above may ripen conditions for upheavals from below, as the Potemkin Mutiny of 1905 did for the Russian revolutions in 1906 and 1917.[57] Additionally, a functional sovereign autocracy may have more stability than democracies because the winning coalition in the former is smaller.[58] Irregular war conditions may also be transported into the state from outside, regardless of the internal levels of voluntarism or coercion.

The conditions for irregular war are pronounced in state collapse combined with foreign intervention. Third parties always intervene out of their own self-interest, but their secondary or declared purposes may include toppling or preserving the incumbent regime, invading and occupying territory, or empowering opposition groups sufficiently to promote defection from the *de jure* political institutions necessary for incumbent state durability. Third-party interventions keep domestic grievances unresolved while fomenting new ones, all while empowering the *de facto* monopolies on force often arising in liminal sovereignty. State constructionism produces conditions favorable for the paralysis or disintegration of the state.[59] In other words, some states experience irregular wars while others do not despite holding many similar characteristics in common, such as location, culture, history, and political systems. The difference rests in whether the state has a functional or dysfunctional sovereign.

Absolutist dictatorships and authoritarian political systems are easier to maintain in states with extractive economic systems, exclusive elites, and economies based around one or few exports, such as oil, sugar, gold, silver, opium, uranium, diamonds, cocaine, cotton, and pepper. Political stability in these states can only remain with extractive economic institutions and the dominance of a minority elite over a majority of people. Productivity, innovation, and liberty are anathema to the forced maintenance of authority. This may explain the onset of state failure commonly following foreign military intervention, as the foreign power often displaces the old authoritarian and selects a leader with a core elite who can produce short-term political stability but who also lack long-term capacity to promote inclusive institutions, rule of law, and justice as fairness. These regimes typically lack centralized political control outside the capital, while elites jealously guard their positions under the protective power of the foreign benefactor. Examples for the Soviets include Mahabad in (1945–6), Azerbaijan (1946–7), and Afghanistan (1979–89). For the U.S., this includes South Vietnam (1955–75), Afghanistan (2001–21), and Iraq (2003–2011). Saudi Arabia and the Emirates served this role in Yemen after 2015, while Turkey, Russia, and the Emirates did the same in Libya after 2012. Russia, Iran, and Turkey also faced similar issues in Syria following the Arab Spring in 2011, though each experienced this in their own way due to the factions each backed and their obligations in the international system. This recurring pattern has intensified around the world since 1945 in the wake of third-party interventions in decolonization, the effects of great power politics, and national self-determination movements.

The hardening of international borders after 1945 occurred alongside an increasingly institutionalized international system emerging out of the decade following the Second World War. This period saw the transition in

Chapter 9. A Dialectic of Irregular War

Europe from empires composed of many nations to states of single or few nations. Rather than negating the Westphalian system, this transformation confirmed and reinforced it, all while bringing new members into its structure.[60] Indeed, this period saw the rapid expansion of the Westphalian system outside of Europe once and for all, dramatically restructuring former European colonies, occupied territories, and directly ruled places especially in the Middle East, Africa, and Asia.

Existing states resist new state formation. Irregular wars are often the violent process through which these parties seek to achieve their competing objectives: the former to maintain the status quo and the latter to change it. Decolonization, independence movements, the Cold War, and shifts in economic development patterns contributed to the formation of new states after 1945, but this did not occur in a vacuum. New states often have great power support and are granted independence to prevent a peer competitor from winning influence in that place. Recent examples include Kosovo, South Sudan, and Timor-Leste in recent times and South Korea, South Vietnam, and South Yemen before that.

State formation occurs in the context of culture, demography, and the environment, while the dynamic factors of opportunity, capacity, and motivation influence the path this process takes.[61] It makes sense then that the U.S. resisted independence for the Philippines between 1898 and 1941, because no other great power was balancing against the U.S. there at the time.[62] The situation changed after the Japanese occupation in 1941, as the U.S. began supporting the Philippine resistance against Japan followed by granting independence in 1946 after nearly half a century of suppressing calls for independence with American military force.[63] Similarly, Portugal's war to retain its colonies in Africa (1961–74) ended not because of a sudden moral *volte-face* in Lisbon, but rather because of great power involvement in the context of the Cold War.[64] That great power involvement involved joint CIA and Israeli support to RENAMO in Mozambique, responsible for over 100,000 civilian deaths and millions of displaced persons, while another joint CIA-Israeli effort directly supported the UNITA insurgency in nearby Angola, resulting in several hundred thousand civilian deaths.[65] Regardless of how the state forms, the people expect it to provide public goods in exchange for the control the sovereign exerts over them.

The sovereign exerts control while extracting rents and labor in exchange for security and other public goods, such as security.[66] Hezbollah and Hamas both provide the public good of security in lieu of a functional sovereign government, thereby increasing their own sovereignty over certain areas relative to competing structures such as the Lebanese government or the Palestinian Authority, respectively. In irregular wars, violence is conducted to redistribute, recognize, reassign, or transform existing arrangements outside of existing political institutions that would normally function to resolve these contradictions. Max Weber emphasized the importance of internal dynamics governing the relationships between the ruler and the ruled within states, while emphasizing consumption over exploitation. John Rawls hinted at a certain harmony between justice and injustice, whereby "In a well-ordered society, one effectively regulated by a shared conception of justice, there is also a public understanding as to what is just and unjust."[67] The legitimacy of the state rests in a harmony of power and force that produces a net positive output on the people affected.[68]

Power and force are two expressions of a state's monopoly on coercion, with the first being granted that authority from the people and the latter demanding that authority.[69] The duration and degree of sovereign force applied against a society can mean the difference between a stable transfer of power, civil war, state failure, or social revolution, as depicted in Figure 11.[70] Authority can be centralized through compliance, participation, and legitimation of the myth of the state.[71] Sovereign legitimacy emanates from the *Weltanschauung* promoted among the ruling elite through the force of coercion or the power of authority. John Locke juxtaposed "the State of War" against "the State of Nature," describing the former as "Force without Authority" and framing rebellion as "an opposition, not to Persons, but Authority."[72] Coercive regimes produce compliance rather than legitimacy, while legitimacy flows from voluntary popular assimilation of the myth and *Weltanschauung* of the ruling class. Since social groups can compete, conflict, or cooperate, power and authority allow some groups to obtain preferential outcomes at the expense of other groups within a political system. Violence occurs when this system breaks down and force is wielded disproportionately to power over time.

States deter, contain, or balance against other states, while proximity, power, and partnership influence their behaviors in war.[73] Since states care most about security for their survival, any challenger to their sovereignty cannot be bargained away. Violence must be the first resort for states and to lose is to cease to exist.[74] States are then compelled to maintain and employ a military institution ready to make violence against other states or risk their own existence. Military institutions are among the most conservative structures a society can produce besides the clergy, and they are often strongly path-dependent even when facing obvious disaster. This structural conservatism is a common theme across much social science literature dealing with power relations in society.[75]

Conventional doctrine allows the state's formal military and security

Chapter 9. A Dialectic of Irregular War

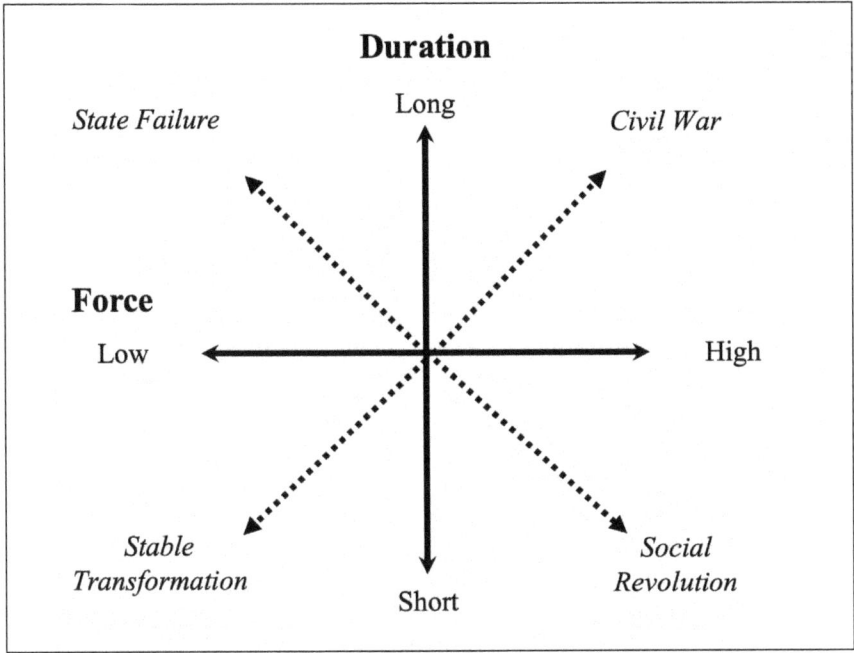

Figure 11. Duration and Force in Power Transitions in the State.

institutions to conduct operations and make plans to preserve or expand state sovereignty. The character of conventional war is framed by *military policy*, which Jomini defined as something influencing a way a state conducts war that is neither politics nor strategy nor tactics.[76] The sequence of war for Clausewitz was to destroy a military, conquer the state, and make peace.[77] It follows that Clausewitz regarded the objectives of offense in war to consist of neutralizing the military threat by any means, conquest of physical territory, and dominance over the opposing state such that the will of the victor is imposed, in that order. The means of destruction and conquest achieve the ends of defeat and capitulation through the application of sovereign violence. The three methods for achieving these ends are invasion, imposing costs, and attrition.

The state is threatened when the outlines of its sovereignty might dissolve and be replaced with new forms and structures. At the same time, states and their institutions resist change, apprehend novelty with suspicion, and defend dysfunction through tradition. Only sufficient force or potent necessity allow for the transformation of sovereignty over social orders, sovereign territories, or political-economic institutions. That transformation may occur in various ways, but through violence it is irregular war.

Irregular Wars and Sovereign Dysfunction

Irregular war is a violent conflict in which a dysfunctional sovereign engages with its own people. Those people are exercising their last option of redressing unmet demands within the social order, incursions or insecurity in the sovereign territory, and exclusion from political-economic institutions. A third-party intervention in such a conflict may exacerbate existing causal factors or create new ones, but without the conflict between people and state there is no irregular war. There are several concepts one may synthesize from the foregoing analysis. First, irregular wars occur through unique anatomies of causes, while the specific milieu of oppression shapes the conduct of the conflict in which belligerents seek to raise legitimacy capital sufficient to claim sovereignty in a perceived zero-sum game begun once a dysfunctional sovereign nudges potential dissatisfaction to kinetic rejection.

Wars between people and the state are far more common and destructive than the conventional wars state militaries are designed to fight. For example, wars within states between 1945 and 1999 caused three times as many casualties, lasted twice as long, and occurred in five times as many countries as conventional wars during the same period.[78] Such conflicts do not have a singular universal origin. Instead, conflicts develop out of complex causes that may involve social groups, identity structures, physical geography, concepts of property, income inequality, regional violence, and so on. The key questions are whether and why a given web of causes is sufficient to transform a dissatisfied but passive society into a mobilized primary group prepared to violently overthrow the dysfunctional sovereign. Additionally, the ends of irregular wars may arrive in forms that take one or all parties by surprise.[79] Irregular wars are often misunderstood by observers, participants, and historians alike unless the *longue durée* of history is considered for each case.

The interactions of the abstract variables described throughout this book were helpful for explaining irregular war as a concept, but each case features a unique anatomy of causes. This is true even when comparing similar cases. In the context of Marxist-Leninist movements, revolution is not an automatic outcome of popular grievances; a mass and a message are necessary conditions, but they are not alone sufficient.[80] Mao Zedong proposed that doctrine, organization, and action were the necessary ingredients for a revolutionary movement.[81] Motivations for transitions to violence are filtered through these causes, exacerbated by contextual social pressures such as fear, shame, justice, and dishonor. The causal factors for the irregular war in Iraq after 2003 included a mixture of a growing population, lagging development, and a chronic lack of meaningful employment.[82] At first glance, these appear different than the three recurring catalysts behind many irregular wars in Latin America, which are sovereignty, foreign occupation, and identity.[83] However, when these are combined with the conditions of general labor conflict, including land tenure disputes, exploitative labor, and repressive political systems, the similarities begin to emerge.[84] Mainstream military theorists often identify religious ideology, social problems, materialism, and nationalism as causes of irregular wars in the developing world.[85] However, the reality is more nuanced, as extractive institutions are linked to state failure while stagnant economies promote starvation and conflict as incentives to participate in the market dwindle.[86] Although the foregoing views have important differences, they demonstrate that the dysfunctional sovereign's interaction with social order, sovereign territory, or political-economic institutions create conditions for irregular war.

A *milieu of oppression* constitutes the combination of unique, path-dependent circumstances surrounding each transition from non-violent resistance to irregular war. The milieu of oppression distinguishes each primary group or insurgency from others. This reality prevents the possibility of crafting a universal history or general theory of the state, war, and the people over time. In other words, *a* human experience cannot describe *the* human experience.[87] Although many have argued that history could end at the feet of the last man, the human experience is instead composed of many histories often contradicting one another regardless of the quantitative picture one paints.[88]

It is within this milieu of oppression that capacity to act and belief in cause interact to promote insurgency and underground support from below. Meanwhile, conscripts and professional soldiers form the conventional forces from above that the state sends to suppress them. This process of action capacity and personal investment in irregular war is depicted visually in Figure 12.[89] Those with a high capacity to act and a high belief in the primary group's cause are the insurgents. The underground supporting the insurgency is composed of those with a high belief in the primary group cause but who have a low capacity to fight. Those with a low belief in the cause but a high capacity to act are the professional soldiers in the conventional forces opposing the insurgency. Those with a low belief in the cause and a low capacity to act, but who find themselves compelled nonetheless to fight for the state, are usually the conscripts or reserves called up to support the professional soldiers against the insurgency.

The impact of those histories upon primary groups grants legitimacy capital to various social ideologies, territorial expressions, and political-economic structures. *Legitimacy capital* is the zero-sum quantity of popular support influencing the outcome of an irregular war. There are several

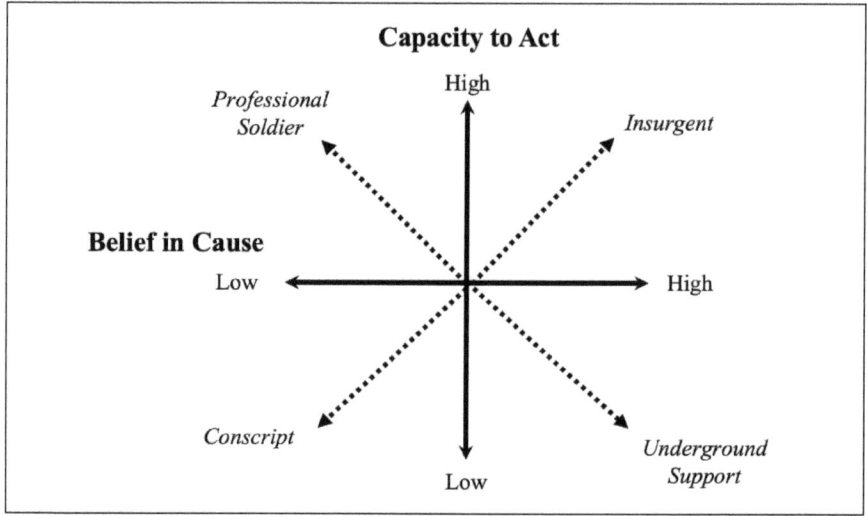

Figure 12. Action Capacity and Personal Investment in Irregular War.

ways to transfer legitimacy capital, including disintegration, elimination, suppression, and exaltation. Disintegration is a result of many factors but is the end of meaningful resistance. Suppression and elimination are the objectives of counterinsurgency operations while mere existence is the triumph of the insurgency.[90] Exaltation occurs when the insurgency triumphs over the incumbent. These ways of transferring, maintaining, or transforming legitimacy capital all depend on how regime rationality interacts with how the legitimacy of its monopoly on force is viewed by those under its dominion. Figure 13 shows the contrast between regime rationality and monopoly on force in terms of legitimacy capital and logic.[91] A totally irrational regime employing totally illegitimate force is most often the preserve of an invading state. Israel's domination of south Lebanon from 1982 to 2000 was likely viewed by many non–Christian Lebanese as both totally illegitimate and totally irrational. In contrast, a totally rational regime employing totally legitimate force constitutes the ideal liberal democracy. Of course, no ideal state exists, but those states with regimes and force approaching rationality and legitimacy are certainly ahead of the game in terms of legitimacy capital. The authoritarian regime has some legitimacy but lacks rationality for those outside the privileged elite. A fourth kind of relationship between regime and force is an illegitimate force employment with a high rationality—this is an insurgency.

The legitimacy of the sovereign, whether functional or not, can be expressed through the effectiveness of its institutions in providing rule of law, administering justice, and executing policy while supporting an

Chapter 9. A Dialectic of Irregular War

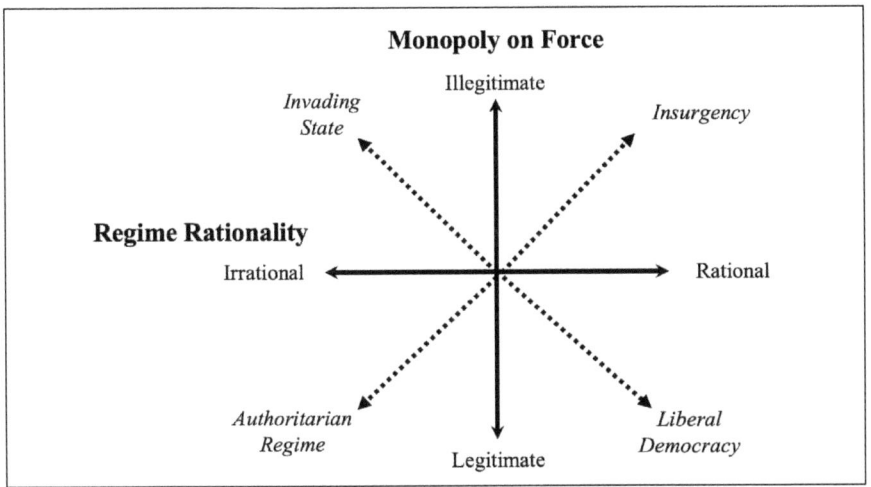

Figure 13. Sovereign Legitimacy and Rationality.

inclusive economic system. Indeed, pluralism and inclusive political institutions are a prerequisite for inclusive economies. These institutions continuously reinforce one another, whether extractive or inclusive.[92] Institutional constraints on participation and incentives to insulate elites work to promote inequality while blocking the majority from working within those institutions to acquire public goods. If institutions effectively constrain elites and include members of society broadly, grievances may be resolved within these institutions and the sovereign is less threatened. If unconstrained, elites either opt to repress the majority, make unenforceable concessions, or invite external intervention.[93] Irregular war flourishes in the liminal space forming from the breakdown in the twin goals of statecraft: appropriation and control.[94]

Irregular war conditions can exist in the context of state or civilizational collapse, as irregular wars display some elements indicative of the collapse of complex societies.[95] These elements include especially decentralization of control, declining enforcement of checks on social behaviors, hoarding resources, and fragmented territorial units incoherently organized and generally uncoordinated. The condition of weak or absent central authority can promote the creation of militias who abrogate established state borders because the sovereignty therein no longer wields sufficient implied force to preclude their necessity and indeed their formation.[96] While the theory of collapse of complex societies asserts that there are no power vacuums left in which states may rush to fill the gaps left by a collapsing state, this applies only to territory *between* states, but not *within* states, and the latter liminal space is precisely where irregular war conditions flourish.[97]

Removing exclusive political institutions and unbalanced social orders can transform the state by reassigning elite status and expanding the franchise. However, slow transformations face strong resistance from existing structures in economic and political institutions, so that changes are more likely during crisis situations and shocks.[98] Failure of political institutions to resolve social issues leads to political violence and "independence by force," since groups can either seek reforms within the system or rebel against it.[99] Ferhat Abbas, the first post-independence president of Algeria, noted in 1954 that "if only the statute had been applied ... which would have permitted the inhabitants to conduct democratically their own affairs, I say that perhaps we should not have had a *maquis* and *maquisards*."[100] Similarly, Zohra Drif, one of the women who planted bombs in the Casbah to begin the Battle of Algiers on September 30, 1956, remarked that she joined the FLN, "an essentially terrorist group," because "France 'had consistently refused the least reform.'"[101] In Vietnam, the Viet Minh exhausted all political recourse to challenge French colonialism through available institutions, including direct appeals to the U.S. and United Nations, before resorting to violence. B.H. Liddell Hart suggested giving the enemy a ladder down which to climb, because a cornered enemy has nothing to lose.[102] An irregular war condition exists when political solutions to institutional problems fail while the means of coercion begin eroding away from the state and into the hands of the primary groups in competition with the sovereign to rectify those problems.

Conflict can arise as a relatively stable but unequal social structure is met by outside influence. Third-party interventions need not consist of conventional forces or states at all, though they often do. Foreign commercial demand for the *grano del oro* (golden grain) of the coffee bean led Central American coffee production to accentuate and accelerate existing socioeconomic inequalities typical in agricultural economies with one or few exports. Some have even argued that the coffee industry was the single most valued process in terms of the resulting exports relative to the low cost of production.[103] Coffee was certainly a powerful force setting conditions for irregular wars in Latin America. Those conditions specifically include labor inequality, rarified wealth systems, and tiny oligarchies running those export commodity markets. Like external influence, disruptions in power concentrations may also contribute to irregular war conditions.

Irregular wars tend to emerge under relatively decentralized power structures in which the authority of the sovereign is uncertain. Diverse forms of social organization are encouraged in the absence of centralized government, while foreign commercial interests act as a solute to keep these entities from binding into a coherent state.[104] The island of New

Guinea is a dramatic example of the interplay between geography and lack of centralized authority upon diversity in language, culture, and even genetics in a relatively small area where 1,300 languages and 40 distinct language families exist, constituting nearly 20 percent of the world's languages in a space the size of California.[105] In Africa, the liminal sovereignty in Somaliland encourages state and non-state political entities to vie with one another to provide security, services, and governance, while in the Middle East the Kurdistan Regional Government exercises *de facto* sovereignty under the shadow of great power politics. The legal aspects of Westphalian borders influence the conventional belligerent's strategy while providing the irregular force an opportunity to exploit those pressures to constrain and impose costs on conventional force basing, overflight, airspace, and support infrastructure.[106] International borders become contentious because of the implications of identity, resources, and communication they entail. An insurgency may ignore those borders altogether in the midst of an irregular war conducted to negate the powers maintaining them.

An insurgency is a violent insurrection within a broader political frame of revolution. Leon Trotsky saw revolution and insurrection as two related concepts, much as Clausewitz saw the relationship between irrational force and the tempering effects of policy. Trotsky considered insurrection to be the action component of the revolution and, according to him, "Its aim was to break down the obstacles which could not be dissolved politically."[107] Examining this connection between revolution and insurrection is important because a significant knowledge gap exists as studies avoid the inextricable link between the degree of function within existing political institutions and the likelihood of armed resistance to the sovereign exercising authority over those institutions.[108] For Trotsky, insurrection was the violent action in the overall revolutionary movement.[109] That violent action is expressed through the conduct of irregular warfare.

Conventional forces often founder in irregular wars because they are bound by the constraints which they invite upon themselves. Amal and Hezbollah in Lebanon illustrate the Westphalian dilemma of limiting confrontation using irregular warfare to appease domestic political and popular aversions to conventional wars, even if the irregular conflict is far bloodier, costlier, and longer in duration.[110] In contrast, irregular forces can draw from their own technological, functional, or structural histories, often generating their own doctrine on the spot, adapting to the burdensome enemy easily by flowing around him like a river over a boulder, though not all irregulars have an easy time at it. Conventional forces seek to maintain equilibrium at any cost, employing legible, duplicative, and predictable methods not because those are effective but because they are easily administered and understood throughout the system employing them.[111]

Irregular war poses a problem of military ethics for conventional forces, as the concept of justice and necessity may have competing roles in such conflicts.[112] Additionally, limiting war with the 1984 Weinberger Doctrine and increasing precision strike capability through technological advancement and so-called information dominance does little to ameliorate the effects of conducting a limited strike that precisely destroys the wrong target, creating strategic and political ramifications far beyond the conflict. Important examples for the U.S. military precisely destroying unintended targets include the airstrike on the Amiriyah bunker in Baghdad that killed 408 civilians in 1991,[113] the airstrike on the Chinese embassy in Serbia that killed three Chinese journalists in 1999, the Mayaguez incident in May 1975, and the U.S. surface-to-air missile strike on Iran Air Flight 655 in 1988 that killed 290 civilians over the Persian Gulf.[114] Irregular war also challenges conventional forces to employ technologies and concepts of operation developed for warfare of a different paradigm.

If the numerically superior, better equipped adversary cannot effectively employ forces or technology then numerical imbalances become meaningless. This occurred for Libya in its war against Tanzania in Uganda in 1979 and later against Chad in 1986, while Syria faced similar challenges against Da'esh from 2014 onward.[115] Mao Zedong recognized this challenge, declaring that "each type of warfare has methods peculiar to itself and methods suitable to regular warfare cannot be applied with success to the special situations that confront guerrillas."[116] New military theories and concepts should be prioritized over new technologies to allow a transcendence beyond traditional military activity.[117] Irregular war is wholly political despite its overtly violent character, and it therefore requires more astute solutions than mere arrangements of forces, employment of expensive weapons, and the use of terms like "kill webs," "grey zone," and "revolutions in military affairs" that all glitter like ornate talismans of the "methodismus" Clausewitz warned against.

Irregular war is a kind of political creative destruction, and as such irregulars often fight in a manner that may appear unconstrained in comparison to conventional forces having limited objectives and clear political constraints. Governments with extractive economies and exclusive political institutions foster the kind of instability, both economic and social, that often degenerates into irregular wars, such as Lebanon's confessional system and Israel's *hafrada* and racially-exclusive Jewish laws. In the latter case, Israel can maintain an artificially strong economy and relative political stability in spite of its oppressive social order, contested territory, and exclusive institutions due to unprecedented support from the U.S.,

Chapter 9. A Dialectic of Irregular War

not unlike periods of history in Japan, South Korea, and the former West Germany. Without this support, it is likely that the artificially functional sovereign Israeli government would devolve into social, territorial, and political-economic conflict characterizing irregular war. Political creative destruction tears the fabric of the state, threatening to form anew its order, boundaries, and institutions.

Insurgency and revolution are often carried beyond their intended threshold of change, partly by the sheer weight of the movement and partly from the traumatic moment in which sovereignty is transformed from potential to kinetic action in attempting to regain uncontested monopoly on legitimate force. In 1880, Friedrich Engels noted this problem in his assessment of the phases of the French Revolution, whereby first the Committee of Public Safety, then the Directory, and finally the Consulate swept away the state, followed later by the continued revolutionary activity in 1830, 1848, and the 1871 Commune.[118] Engels placed supreme importance upon revolutionary social change sprouting from the Protestant Reformation, the 1525 Peasant's War, and finally from Calvinism's triumph in the foundation of the Dutch Republic in 1581. This kind of deep historical analysis is lacking in current studies and yet it reveals much about the impact of a dysfunctional sovereign over time. The revolutionary activity in nineteenth century France was as much about the dysfunctional sovereign as it was about the people, and states would be remiss to exclude this fact from strategic planning in irregular warfare.

It is naïve to frame the objectives in irregular wars so simply as contests over popular favor, though the population is certainly important. The conflict does include a competition over individuals who may either collaborate, abstain, or defect. Popular support is crucial in warfare, regardless of its character or the form of government executing it. Niccolò Machiavelli noticed this from his own experience leading troops in Italy, observing in 1531 that "for always, no matter how powerful one's armies, in order to enter a country, one needs the goodwill of the inhabitants."[119] If this goodwill was lacking, however, Machiavelli expected that each citizen would become an enemy against the state, quipping that "the best fortress that exists is to avoid being hated by the people."[120] Hate is a powerful force for resistance, a force Jean-Paul Sartre likened to a magnetic field.[121]

All of this is to say that there exists a continuum along which people either totally resist or completely collaborate with the insurgency. This continuum of civilian control between insurgencies and states is depicted in Figure 14.[122] The insurgent group would understand an adversary population to consist of those who are proactively resisting that insurgency. On the other side, an insurgent group would understand voluntary collaborators to be an allied population. Those civilians who are protectively

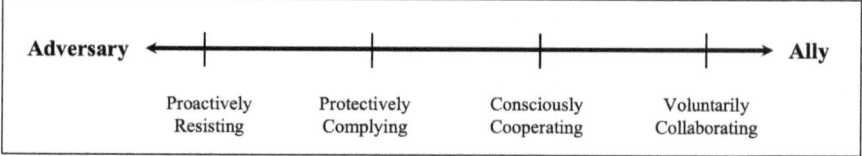

Figure 14. Insurgent-Civilian Control Continuum.

complying with the insurgency are complying contingently to protect something, usually their families, livelihoods, or their own lives. In contrast, those civilians who are consciously cooperating understand that their assistance actively supports the insurgency, yet this may still be based on a rational calculation about survival and could also be affected by proximity bias. The control of the civilian population is not simply a matter of supporters and detractors; instead, the relationship is far more complex and is based on a web of often competing motivations.

Irregular war also includes the related contests over resources, territory, narratives, and other issues influencing one another to express, promote, and defend interests in social orders and political institutions. States, civil wars, and state formation should be viewed in terms of arrangements of dominant coalitions and political settlements rather than as a functional progression to an ideal stability that no real state possesses.[123] State approaches to irregular wars focusing on disarmament, demobilization, and reintegration into the social and political structure during and after a conflict may be more effective than those seeking to destroy or defeat the insurgency. Such a strategy can only be formed after the causes of the insurgency are fully grasped and security sectors of the incumbent sovereign are sufficiently productive for its constituents.[124] Furthermore, security sector reform is the most effective approach to countering insurgents in post-colonial states, as opposed to the conventional deployment of military forces to conduct what are essentially colonial policing operations.[125] This failure to address root causes of insurgency at the national level was evident as the U.S. preferred to use special operations forces to conduct direct action missions rather than foreign internal defense or security forces assistance, which led to the failure of the Civilian Irregular Defense Group concept by 1962 in Vietnam under Operation Switchback.[126] Despite this, security sector reform has been an important focus area in stability operations, as it seeks to integrate militias into the state through civil and military security institutions rather than simply suppressing insurgency.[127]

The outcome of an irregular war depends on several factors, regardless of whether states reform security sectors, conduct unconventional warfare, or violently suppress an insurgency. Among others, these factors include strategies employed during the conduct of irregular warfare, whether the state understood the insurgency, and how the belligerents decided to address the systemic problems leading to the insurgency in the first place. Likewise, third-party decisions leading to intervention generally include views of the conflict that are more aspirational than predictive of outcomes in irregular warfare and the attendant repercussions among the international community.

Irregular war involves a tension over unity or separation, regardless of the structural changes to social orders in revolution, secession resulting from division of the sovereign territory, or political changes through dissolution of institutions. In some complex insurgencies, all of these occur simultaneously, such as in Iraq, Kosovo, Sudan, and the Caucasus. A central question in irregular war is whether irregulars and the state should unite or separate, and it is a question resolved in violence. The dysfunctional sovereign may be preserved, transformed, or replaced to correct the antagonisms leading to violence. Whether this occurs or not, violence may end if the insurgents agree to reconciliation or federation with the sovereign, seek and gain autonomy from the sovereign, or succeed in separation from the sovereign territory. Any other outcome is artificial as there can be no natural *status quo ante bellum* between conflicting elements of a polity violently contesting the overall order, control, and productivity of the state. This artificiality may lead to some observers claiming their revolution is perpetual, but this is a myth.

Perpetual revolutions require a mythos of an undying enemy. For economic-structural revolutions like those under Marxism-Leninism, the enemy is capitalism. For religious-cultural revolutions like Iran, the enemy is *gharbzadehgi* ("Westoxification"). This perpetuity justifies radical centralized action not otherwise socially palatable in the absence of real or invented dangers of the abstract enemy. This is a convenient tool to justify the actions in revolutionary governments that were born out of irregular wars but that have become wholly conventional, traditional, and authoritarian as they embark on their own races to build powerful conventional forces.

Conventional forces will continue facing these same challenges in irregular wars, challenges experienced since the Peace of Westphalia brought these fighting units to the center of states in 1648. Irregular war is not a matter of technology, tactics, and territory, though conventional war certainly is. French *para* Roger Trinquier continued to believe that the sole reason France lost Algeria and Indochina was a failure to employ enough resources properly, not that France misunderstood the enemy, even after witnessing France's failures in both places firsthand.[128] Entertaining structural and functional changes to conventional forces in irregular wars was anathema to Trinquier's conception of the nature of war and his place in it. Trinquier believed that "once we have occupied the terrain, one ought to have the will and the patience to track [an insurgent] down until we have annihilated him."[129] For this, he failed in every conflict. Trinquier's legacy pulses through U.S. counterinsurgency and irregular warfare doctrine, even after numerous attempts to replicate the French approach have ended in ashes, death, and national humiliation from Afghanistan to Vietnam and beyond. As long as states continue viewing irregular war as a lesser included capability of conventional forces, those forces, and indeed those states, will continue failing.

A dialectic of irregular war involves an insurgency arising from some anatomy of causes against its antithesis, the incumbent sovereign. That sovereign is dysfunctional as it fails to meet expectations in stabilizing social orders, securing sovereign territories, and including citizens in the political-economic institutions producing public goods. As the monopoly on legitimate coercion becomes contested, power devolves to force and a liminal sovereignty opens. In this gap, third parties intervene, armed factions organize to resist the state, and violence works to synthesize a new sovereign, one ideally functioning to meet the people's needs. This dialectic occurs in the framework of social causes, is conducted as irregular warfare, and features various outcomes that are difficult to predict at the outset of hostilities.

The onset, conduct, and outcomes of irregular wars are more complex than conventional wars because the sources of dysfunction, violence, and reconciliation often originate from disparate sources that taken alone would not create such conditions. Understanding this complexity requires focused analysis of each case to identify the anatomies of causes, milieux of oppression, and structures of legitimacy capital in the liminal space between the collapse of a dysfunctional sovereign and the popular movement mobilized to rectify problems in the social order, sovereign territory, or political-economic institutions of the wilting state.

Concluding Irregular Wars

> In reading such facts, we must draw from them not rules, but hints; for what has been done once may be done again.—Antoine-Henri Jomini[1]

The conditions of order, control, and productivity are central to an inquiry into irregular war. This study has shown that an ideally integrated social order, controlled sovereign territory, and fair political-economic institutions can devolve into disintegrated, uncontrolled, and unproductive structures given the introduction of a sufficiently dysfunctional sovereign. The anatomy of events leading the dysfunctional sovereign to negatively impact these conditions can be studied, analyzed, and explained using the conceptual framework presented in this study. Irregular war conditions in specific cases may include economic shocks, widespread resource scarcity, suppression, third-party intervention, and proximity to other conflicts, but the existing state is threatened most when these and other conditions are combined with a sovereign that cannot properly function to provide escape valves for its citizens through which to seek change. Those citizens may opt for collective action against the dysfunctional sovereign, ranging from strikes and protests to secession and revolution.

It is because of the centrality of the individual in irregular wars as both thinker and actor that this book orients the analysis on the causes transforming a citizen to an insurgent, rather than focusing on the conduct or effects of irregular wars. Irregular war is not inevitable in those circumstances in which sovereign dysfunction interacts with social orders, sovereign territories, and political-economic institutions; instead, the individual having the potential to resist must receive a signal to act that communicates forcefully enough that a critical mass of others will act, or have acted, or else none will act.[2] The fall of the Batista government in Cuba is an example of what could be called "explosively congruent expectation,"[3] whereby a critical mass of potential resistance elements interpreted something in the *zeitgeist* as a signal to act.

My approach to the problem consciously shifts attention from the state to the violent factions arrayed against it, facilitating a superior understanding of the outcomes for each participant, whether they are insurgents, states, or third parties. Such analysis implies that there are four outcomes in irregular wars, depending on which actor is considered. Insurgents may whither and retreat such as in Mahabad in 1946, reach a stalemate such as in the Sahrawi Arab Democratic Republic in 1991, integrate as in Yemen in 1990, or disintegrate as happened with elements of the Former Republic of Yugoslavia from 1993 to 2001. Meanwhile, state outcomes include loss, gain, settlement, or resignation. Third parties may seek balancing, stability, guarantees, or change of status. For all parties, these outcomes must be understood in terms of who is fighting, what they are fighting for, and what they seek to gain through violence.

Most states start wars assuming a swift victory.[4] Irregular wars offer no such outcomes. The words of Yakup Şevki, a leader in the Turkish War of Independence (1919–23), still ring true: "History has shown many times that resistance movements based on legitimate rights can force even strong states to change their aim and purpose."[5] And these movements are far more frequent than conventional wars, despite the relatively tiny amount of scholarship devoted to these in contrast to conventional war studies.[6] Most organized violent conflicts since 1648 included at least one non-state group participating, while most wars fought since 1946 were between a state and an organized violent group that was not a state. Although scholars agree that "history is littered with civil wars,"[7] wars between states dominate the scholarly focus of organized armed conflicts. Not coincidentally, these types of internal political instability have constituted most examples of war since 1946 even while contemporary scholarship continues framing wars overwhelmingly as conventional conflicts between states, or at least derivatives of those conventional conflicts.

Studies of irregular wars should include an analysis of causes, support, and violence in the promising areas for research in irregular wars, with particular focus on civil wars.[8] These areas focus on the individual, the specific environment, and the logic of violence. The diversity in research designs and methods in this area indicate it is yet a young field.[9] Likewise, insurgency and revolution are part of the "big questions" in political science that are lacking theoretical treatments due to "inattention to basic research design,"[10] an issue political scientists have long acknowledged but have yet to fix.[11] The problematic state of theory in war and international politics offers hope for future studies of irregular wars because the field has yet to grasp the underlying factors that generally underpin the interactions between people and states.[12]

Anthropologists and sociologists over the previous century have promoted various arguments to explain the collapse of complex societies. These arguments tend to fall under the general theme of complex societies collapsing under the weight of the dysfunctional sovereign unable to productively interact with social orders and political-economic institutions.[13] These circumstances require problem-solving strategies, i.e., solutions. Failing to solve the problems adequately can lead to collapse. There is no military *solution* to irregular war; there may only be a military *component* to a larger strategy that is framed in political ends, flanked by all instruments of national power, working in concert through the direction of a single, unified leadership and grand strategy.

Irregular war outcomes are based on the interaction between a state of hostility and a state of expectation. In this interaction, decision and satisfaction govern whether the belligerents remain engaged in violent conflict. Figure 15 depicts these outcomes, drawing from the ends of war that Carl von Clausewitz first described in the early nineteenth century.[14] When belligerents are sufficiently satisfied with the situation and the hostilities are clearly ended, there is a decisive end to the irregular war. This ideal outcome is quite elusive but has occurred in some situations, such as the end of the American Revolutionary War in 1783, the end of the Indonesian National Revolution in 1949, and the Israel-Hezbollah War in 2006. One common element these and similar examples have is the decisive end of hostilities being enshrined through internationally recognized instruments such as treaties or resolutions. When the situation is stable but the hostilities are not clearly finished, an irregular war often develops into a war of attrition using a strategy to erode will to fight. In this situation, such as the stalemate between Da'esh and the U.S.-backed coalition in 2015 and early 2016 in Iraq, the belligerents favor the use of overwhelming force, degradation of adversary capability, and an exploitation of the adversary's will to fight. When the hostilities never clearly finish and both sides remain totally unsatisfied with the situation, a frozen conflict occurs. This outcome is exemplified by Israel's military occupation of Palestine's West Bank and continued conflict against Gaza, as well as in other places like South Ossetia, Transnistria, Nagorno-Karabakh, and Kashmir. Finally, when one side surrenders or agrees to a punitive treaty, the sides are not mutually satisfied but the hostilities are clearly over. Examples include the Huks evaporating in the Philippines after the Second World War, Kosovo achieving independence in 2008 despite Serbian objections and resistance, and the end of the Algerian War of Independence in 1962.

Irregular wars are violent pursuits of justice, fairness, and liberty. It is only by employing a theory avoiding problems of scope, method, bias, and character that one may begin to understand and explain the structures of conflict leading to insurgency and revolution, not only to prepare for these

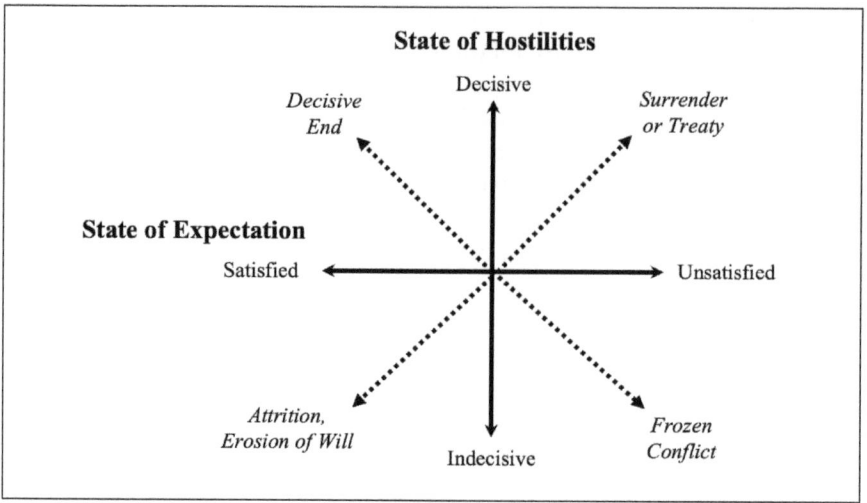

Figure 15. Irregular War Outcomes.

inevitable irregular wars, but perhaps even to prevent some. Before the problem of violence can be solved, a theory of a particular irregular war must explain the various conditions leading to the conflict before the last ballot is cast in deference to the first bullet fired. The theory must address the path dependence for all belligerents in the onset and conduct of war, whether an insurgency, conventional force, or third party. To that end, the framework of irregular war I have developed in this book incorporates the *longue durée* of history, the power of contingency, and a narrative analysis of each case to offer a new way of theorizing about the conditions often leading to organized political violence against states. Every absolute power system will eventually be absolutely torn apart as those silenced under the old regime wrest their voices from the throat of the perishing sovereign.

Chapter Notes

Chapter 1

1. Scott (1998, 206).
2. Examples are Weber (1918); Waltz (1959); Tilly (1990); and Wagner (2007).
3. James (2006, 17).
4. U.S. military doctrine defines unconventional warfare as "Operations conducted by, with, or through irregular forces in support of a resistance movement, an insurgency, or conventional military operations" (U.S. Army Field Manual 3-05.201 2007), while defining irregular warfare as "a violent struggle among state and nonstate actors for legitimacy and influence over the relevant populations" (Joint Publication 1-02 2014). Notably, neither of these definitions applies to irregular *war*—instead, it is only the conduct, *warfare*, that the military seeks to define. U.S. Army Operating Forces doctrine also places all unconventional warfare forces in offensive operations only; there is no defensive role for unconventional forces in U.S. military doctrine (Larsen and Wade 2019, 4–18; U.S. Army Field Manual 3-05.130 2008, 1–2; Joint Publication 3-05 2014, II-2).
5. Those three are modernization theory, Marxist theory, and state-centered analyses. See Goodwin (2005, 405) for more on these three.
6. Those six are power resources theory, Marxist theories of state, critical theory, emancipatory theory, globalization theory, and world systems theory. See Van den Berg and Janoski (2005, 78–83) for a discussion of these.
7. Goodwin (2005, 407).
8. Darwin, quoted in UK Ministry of Defence *Red Teaming Handbook*, 3rd ed. (2021, x).
9. For example, conflict theories developed in the twentieth century commonly were divided diametrically into either Marxist or non-Marxist constructs, whereby the latter lacked the progressive social determinism of the former (Van den Berg and Janoski 2005, 72–95). Some have argued that world systems theory and globalization theory are simply conflict theories at different levels of analysis, with some theories employing the core-periphery conceptualization of the international system in lieu of the master-slave model for citizens of the state (Van den Berg and Janoski 2005, 82). In contrast, more recent scholarship has argued that no grand theory of the interaction between regime change and contentious politics could exist because the outcomes are contingent and path-dependent (Tilly 2005, 423). For more on these frameworks, see Geddes (2009) and King, Keohane, and Verba (1994). See Hegel (1807) for the original master-slave argument.
10. Examples of the social order theme include Arendt (1963); Skocpol (1979); Goodwin (2001); Kalyvas (2006); and Acemoglu and Robinson (2012a).
11. Examples of the state-centered approach include Huntington (1957); Waltz (1959; 1979); and Walt (1996).
12. Examples of the military instrument approach include Clausewitz (1832); von der Heydte (1972); Waltzer (1977); Kilcullen (2009); Jones (2017); and Biddle (2021).
13. Skocpol (1979).
14. See Parsa (2000); Amanat (2017); and Acemoglu and Robinson (2019). France, China, and the U.S. were examined in a similar manner in Fukuyama (2012; 2015).

Notes—Chapter 1

15. Waltz (1959).
16. Waltz (1979).
17. E.g., Walt (1996).
18. Shapers of this model include Waltz (1979); Huntington (1968); Morgenthau (1948); and Lipset (1959).
19. Acemoglu and Robinson (2012a).
20. The authors did expand their aperture in later works, e.g., Acemoglu and Robinson (2019), but they retained the narrow character of their analysis at the expense of a broader consideration of causal factors. It will be helpful to engage with their conclusions while looking more specifically at the role of violence when the state is not functioning as expected.
21. Arendt (1963).
22. Prominent examples include Goodwin (2001); Kalyvas (1999; 2001; 2006); Kalyvas and Kocher (2007; 2009); and Jentzsch, Kalyvas, and Schubiger (2015).
23. Goodwin (2001).
24. Kalyvas (1999; 2006).
25. Huntington (1957).
26. Many examples of the reprint tendency exist, especially the reprints of the 1940 *Small Wars Manual*, the 1921 *Advanced Base Operations in Micronesia* by Earl Ellis, and the 1922 "The Evolution of a Revolt" by T.E. Lawrence. The reinvention of old ideas, often unwittingly, is also very common. A recent example is the "new" *Resistance Operating Concept* that U.S. Special Operations Forces bestowed upon Ukrainian irregulars after Russia's invasion in 2022 (Liebermann 2022). The *Resistance Operating Concept* (Fiala 2020) was developed after 2013 at U.S. Army Special Operations Command, yet it follows the same ideas in Von Dach's 1957 *Total Resistance* and von der Heydte's 1972 *Modern Irregular Warfare as a Phenomenon of Military Policy*. For more information, see ATP 3-18.20 (S//NF) Advanced Special Operations Techniques (ASOT) (U).
27. Representative examples include Jones (2017) and Biddle (2021).
28. E.g., Rogers (2016).
29. Kilcullen (2009).
30. These include especially Roger Trinquer, Robert Taber, David Galula, John Nagl, and Max Boot.
31. Prominent authors committing this error include Kilcullen (2009); Arquilla (2011); Jones (2017); Pollack (2018); and Biddle (2021).
32. Biddle (2021) in the first instance, Pollack (2018) in the second.
33. Mearsheimer (2001, 6–9).
34. Perloff (2014, 52).
35. Perloff (2014, 71).
36. The type of data to be collected called for a qualitative comparative approach that could analyze cases with fuzzy data falling between large-N and small-N sets. The construction of theory in this sense refers to the "attempt to make sense of the world through generalizations of empirical phenomena" (Timmermans and Tavory 2012, 182). See Hunter (2005) and Fernández, et al. (2007) for discussions of how these methods can complement one another.
37. See Glaser and Strauss (1967) for the seminal work on Grounded Theory.
38. Howell (2015).
39. Rawls (1972); Shariati (1978). See Amanat (2017, 695–99) for a biographical sketch of Ali Shariati.
40. Jasper (2005, 132).
41. For more on narrative methods, see Geddes (2009, 131–73); Levi and Weingast (2016); and Bates, et al. (2000). For more on descriptive inference, see King, Keohane, and Verba (1994, 34–74).
42. E.g., Alden, Thakur, and Arnold (2011).
43. Pollack (2019).
44. Crossett (2012); Tompkins (2013).
45. King, Keohane, and Verba (1994, 8).
46. See, e.g., Morgenthau (1948); Waltz (1959); Moore (1967); Skocpol (1979); Waltz (1979); McNeill (1982); Tilly (1990); and Fukuyama (2012).
47. Jones (2017, 5). Despite these dismal success rates for secession, the U.S. has a long history of supporting secessionist states. A minority were successful in breaking away from their parent state: Texas in 1848, Kosovo in 2008, and South Sudan in 2011. Russia has had a similar affinity for this low probability of success activity, in places including South Ossetia, Abkhazia, Nagorno-Karabagh, Transnistria, Donetsk People's Republic, and Luhansk People's Republic, among others.
48. Levy and Thompson (2010, 186).
49. Bassiouni (2008, 712). See also Parry (2015) for a perspective on *jus bellum justum* theory in the context of irregular war, building upon the ideas Waltzer (1977) presented on the topic for conventional forces.

50. Connable and Libicki (2010).
51. Martin and Nordstrom (1992, 3).
52. Van Creveld (1990, 22). Kalyvas (2006, 54) identified 146 civil wars during this same period.
53. Schubert (2013).
54. Alden, Thakur, and Arnold (2011, 17).
55. Eagleton (1963); cf. Weigert (1963).
56. Simatupang (1972).
57. LeRichie and Arnold (2012).
58. Clausewitz (1832).
59. Scott's (1998) important work on this problem is essential for understanding the folly of such a worldview.
60. Ibn Khaldun's *Muqaddimah*, first available in 1377 CE, was arguably the first work of sociology in any language. Celebrated historian Arnold Toynbee worshipfully described the book as "undoubtedly the greatest work of its kind that has ever yet been created by any mind in any time or place" (quoted in the introduction to Khaldun 1377, viii).

Chapter 2

1. Kalyvas (2006, 218).
2. Rawls (1971, 47).
3. Bronner (2017, 7).
4. See Thompson (2013, 275–308) for a comparative analysis of writings by Ali Shariati and Sayyid Qutb.
5. Weber (1918); Durkheim (1893); and Shariati (1970, 57–8).
6. Durkheim (1912, 155).
7. Milgram (1963, 168).
8. Adapted from Shariati (1979, 44).
9. See Amanat (2017, 690–5) for an objective overview of Al-e Ahmad's development of the concept of *gharbzadehgi*.
10. It is worthwhile to note here a representative example of scholars misreading and misunderstanding sources external to their oft-cited repertoires of Western sources, with direct implications for irregular wars. In his influential work *The Transformation of War*, Martin van Creveld (1991) laid bare his poor grasp of non-Western sources in mentioning the Persian concept of *gharbzadehgi* in passing, failing to address the significance of this and similar concepts to the belligerents in the conflicts he sought to explain. Nor did he discuss the centrality of this idea to Jalal Al-e Ahmad or Ali Shariati and the revolutions that followed, or similar ideas for Arab nationalism that Edward Said (1978) examined. Van Creveld then went on to warn Western militaries about a reified Islamic extremism, failing to explain why some Muslim groups have, as he claimed, "no difficulty finding volunteers ready to commit suicide for [Islam's] sake" (van Creveld 1991, 142). Needless to say, reality is more nuanced.
11. Waltz (1959).
12. Waltz (1959, 225).
13. Weber (1918, 28).
14. Jung (1957, 16).
15. Mahdi (1987, 210–3).
16. Said (1978, 206).
17. Shariati (1970, 19).
18. Shariati (1970, 57, 73).
19. See Zarrinkoub (1957) for a different perspective on authority in Persian culture.
20. Goldstone (2014).
21. Brafman and Beckstrom (2006, 99).
22. Quoted in Taber (1965, 59).
23. Mao (1937, 44).
24. Simatupang (1972, 116).
25. "Dwitunggal" means two languages of power, literally "bilingual" (Simatupang 1972, 212).
26. Eagleton (1963, 50–3).
27. Adapted from the frameworks given in Rawls (1971, 66, 266).
28. Rawls (1971, 49).
29. Rawls (1971, 8).
30. Kant (1781); Hobbes (1651); and Hegel (1820; 1837).
31. Rawls (1971, 15).
32. Fukuyama (2007, 4).
33. For further reading, see Olson (1965); Schelling (1960; 1978); and Geddes (2009, 195–8, 203).
34. Schelling (1978, 76).
35. Schelling (1960, 32).
36. See Tilly (2003, 58) for further discussion of factions participating in irregular wars, both for and against the state.
37. Tilly (2003, 104).
38. Rawls (1971, 5).
39. Cf. the variety of ideas about the characteristics of the state in Tainter (1988, 26–7) and Morgenthau (1948).
40. Locke (1690, 428).
41. Locke (1690, 406–28).
42. See Acemoglu and Robinson (2012a,

216) and Goodwin (2001) for discussions of these policy effects in general contexts.
43. Rawls (1971, 47).
44. Rawls (1971, 6).
45. Locke (1690, 406).
46. Rousseau (1762, 61), emphasis in original.
47. Rousseau (1762, 69).
48. Hobbes (1651, 627).
49. Jefferson (1776).
50. Rousseau (1762, 49).
51. More (1516, 145). See also Erasmus (1512) for a related discussion of the roles of law and sovereignty in society.
52. Locke (1690, 351, 417).
53. Acemoglu and Robinson (2019, 18, 64).
54. Scott (1985, 38).
55. Tainter (1988, 170–1).
56. Tainter (1988, 169–71, 175).
57. Von Clausewitz (1832, 537).
58. Von Clausewitz (1832, 537). Prominent echoes of this oversight can be found in van Creveld (1991); Nagl (2005); Kilcullen (2009); and Jones (2017) among many others.
59. Byman (2007).
60. Zimbardo (2007, vii).
61. Forsythe (2010, 12–14).
62. Perloff (2014, 125). The study of attitudes grew in the 1930s, representing a new direction in social science research beyond the origins of the field laid by Marx, Durkheim, and Weber. See Perloff (2014) for an in-depth analysis of attitudes and society.
63. Perloff (2014, 49, 228).
64. This is detailed in Tuckman (1965) and Tuckman and Jensen (1977).
65. Forsythe (2010, 130).
66. Keohane (1984, 65). In contrast to Keohane, Spears, et al. (2001) argued self-categorization theory demonstrates that "responsiveness to a group norm is not a mindless or irrational process reflecting a reduced sense of self ... but may be a conscious and rational process relating to a meaningful sense of identity" (336).
67. Von Dach (1957, vi).
68. See Acemoglu and Robinson (2019) for a discussion of 'asabiyyah in a political science context.
69. See also Hovsepian (2008, 44).
70. Brubaker (2006). Others include Didier and Aoun (2019, v) and Pollack (2019, 488).
71. Zarrinkoub (1957, 65).
72. Mao (1937, 92). See the "eight rules" and "three remarks" in Mao (1937) for the details of his 'asabiyyah.
73. Rodriguez (2009).
74. Rodriguez (2009, 109).
75. *Anomie* is a detachment from the group-regulated system of norms. Anomie pushes individuals out of a group, and in the extreme case these individuals may lose all attachment to any group identity, suffering from what Durkheim diagnosed as "the malady of the infinite." On anomie, see Durkheim (1893). See Didier and Aoun (2019) for a recent, albeit flawed, use of anomie in studying insurgent groups.
76. Based on concepts in Durkheim (1912); Zarrinkoub (1957, 65); Gellner (1988, 239); Pollack (2019, 488); and Ibn Khaldun (1375).
77. Pew (2015). Christians, Zoroastrians, Jews, Druze, and others are also monotheistic, but the term *tawhid* is more precise in the context of Islam and its meaning to the world's 2 billion Muslims.

Chapter 3

1. Schelling (1960, 3).
2. Jomini (1862, 126).
3. Clausewitz (1832, 79), emphasis in original.
4. Clausewitz (1832, 96).
5. Clausewitz (1832, 199).
6. Clausewitz (1832, 534).
7. Jomini (1862, 57); Clausewitz (1832, 512).
8. See Schmitt (1932, 54). Later, Lenin's *Der Partisankampf* [Partisan War] was published 30 September 1906 and was among the earliest attempts to describe the partisan as an independent irregular force rather than an auxiliary attachment to a national and global narrative, poised in stark contrast to state power and the Westphalian political system. For a view sympathetic to the state, see Waltz (1959) and Kissinger (2015). See Schmitt (1932, 34) for a view sympathetic to the partisan.
9. Van Creveld (1991, 64–5).
10. Some of these traditionalist views include Mahan (1890) and Liddell Hart (1954).
11. Van Creveld (1991, 63); cf. Kalyvas (2006).

Notes—Chapter 3

12. Krieg and Rickli (2019, 37, 42) noted that van Creveld's work "constitutes the baseline in Western thinking on war against which both the nature and character of war is evaluated."
13. See, e.g., Krieg and Rickli (2019, 10).
14. Liddell Hart (1954, 324).
15. In contrast, Fukuyama (2012; 2015) studied the origins of politics and war in Asia alongside Europe, while Tilly (1990) and McNeill (1982) provided histories of the state and war in Europe without claiming these to be universal, an error van Creveld was not shy to make. The analyses in Acemoglu and Robinson (2012a) and Scott (2017) are superior, not only in their case study selection but also in their findings.
16. This is the most common focus in Western scholarship on military theory, history, and strategy, with powerful voices including Waltzer (1977) and Mahan (1890) exerting a magnetic effect on the direction of such scholarship.
17. Mao (1937, 49); Clausewitz (1832); and Lenin (1906).
18. Byely, et al. (1972), quoted in Jonsson (2019, 3).
19. These include Arquilla (2011) and Nagl (2005, 25).
20. Nagl (2005) committed this error. See Melson (2019) for an exceptional study of Prussian and German military theory from 1800 to 1945. See Fravel (2019) for an equally important study of the development of Chinese military strategy from 1949 to 2019.
21. Fravel (2019). This primary strategy of the Chinese People's Liberation Army is not unlike that which great Phoenician general Hannibal Barca used at Cannae in 216 BCE to lure the superior Roman legions into his weak center prior to utterly annihilating a fifth of the Roman adult male population in a single day through a masterful double envelopment. The battle at Cannae produced more combat deaths than on any other day in human history (Hackett 2018).
22. Berger (2020, 4).
23. Gray (1989); Krulak (1997).
24. Huntington (1957, 263).
25. Speier (1941, 445).
26. Bueno de Mesquita, et al. (2003, 405).
27. Fearon (1995, 380).
28. Fearon (1995, 408); Mearsheimer (2001, 337).
29. See Jomini (1862, 8–26) for the list of nine motivations for states to go to war.
30. Nagl (2005). Taber (1965) was an early proponent of this view.
31. See also Engels (1880); Lenin (1918, 157); Stalin (1938, 8); Arendt (1963, 67); and Luttwak (1989, 241).
32. One prominent example is Galula (1964).
33. Gray (2006, iv).
34. Gray (2006, iv). See Cleveland and Egel (2020) for a study of an American way of irregular war.
35. Quoted in Gray (2006, 2).
36. Arendt (1963, 6).
37. Arendt (1963, 9).
38. Jameson quoted in Rodriguez (2009, 2).
39. Inkles and Smith (1974), but cf. Pollack (2019, 242) for a different perspective.
40. Geertz (1973, 56).
41. Keohane (1984).
42. Uncertainty is felt by all participants during irregular wars, even those contributing significantly to planning the uprising. In the Kurdish Self Administration of Northeast Syria, one observer employed a useful metaphor: "The revolution in Rojava is like a newborn child—we don't know how it will grow up" (Knapp, et al. 2016, 58).
43. Liddel Hart (1954, 369).
44. Kilcullen (2009).
45. See, e.g., McNeill (1982).
46. Liddell Hart (1954, xv).
47. Liddell Hart (1954, 27); Biddle (2021); but cf. Hackett (2018, 47).
48. Quoted in Barr (2011, 115).
49. Liddell Hart (1954, 370).
50. Liddell Hart (1954, 370).
51. Levy and Thompson (2010).
52. Arquilla (2011, 3–5).
53. Tainter (1988, 3).
54. Arendt (1963, 106); Acemoglu and Robinson (2019).
55. Arendt (1963, 47).
56. Schell quoted in Arendt (1963, xv).
57. Le Bon (1895).
58. Le Bon (1895, 3), but cf. Shariati (1979) and Nietzsche (1911, 203–4).
59. Kropotkin (1892, 63).
60. Hegel (1807).
61. See Byrd and Miri (2018, 58) for further discussion of Shariati's views on the population and mass movements.
62. Atabaki and Zürcher (2017, 5).

63. Forsythe (2010, 503).
64. Arendt (1963, 262).
65. Acemoglu and Robinson (2012, 43); Keohane (1984). Spears, et al. (2001) noted that "responsiveness to a group norm is not a mindless or irrational process reflecting a reduced sense of self ... but may be a conscious and rational process relating to a meaningful sense of identity" (336).
66. Rogers (2016).
67. Goldstone (2014, 11–12). Examples of the erring view include Le Bon (1892) and Kilcullen (2009).
68. Acemoglu and Robinson (2012a, 443); Lipset (1959). Samuel Huntington's *Political Order and Changing Societies* (1968) provided another important refutation of modernization theory. In the preface to the 2006 edition, Francis Fukuyama noted the importance of Huntington's challenge to Lipset (1959).
69. One example is Huntington's (1968) introduction of the concept of praetorianism, which he defined as the reaction of elites when social changes outpace political and economic institutions.
70. Schumpeter (1919; 1942). Nearly a century before Schumpeter's (1942) theory of economic creative destruction, Russian revolutionary agitator Mikhail Bukharin proclaimed: "Destruction is also creative" (quoted in Serge 1951, 453). For various arguments on the classification of governments, see Acemoglu and Robinson (2012a); Levitsky and Way (2010); Keohane (1984); Svolik (2012); Moore (1967); Wittfogel (1957); Waltz (1979); Mearsheimer (2011); and Arendt (1951).
71. See, e.g., Walt (1996) and Huntington (1957).
72. Osanka (1962); Kalyvas (2006); Ismail (2018); Tilly (1990); Acemoglu and Robinson (2012b); and Martin and Nordstrom (1992).
73. Acemoglu and Robinson (2012b, 42).
74. Skocpol (1979, 172). Geddes (2009) questioned Skocpol's fundamental arguments, asking, "Are revolutions infrequent because of the absence of appropriate structural conditions ... or because foreign threats have less causal impact than Skocpol believes?" (108). Geddes attributed this problem to Skocpol selecting cases on the dependent variable of revolutionary situations, rather than also including situations in which revolutions did not occur or were unsuccessful.
75. Mahoney (2010, 129).
76. Kropotkin (1892, 209).
77. Chehab (2005, 239).
78. Pollack (2019); Huntington (1993). This is a favorite worldview for American neoconservatives, some of whom are also influential military theorists and policymakers.
79. North, Wallis, and Weingast (2013, 13, 27).
80. Scott (1998) better captured the earliest social orders, while Acemoglu and Robinson (2012a) provided a more realistic explanation for social orders and the political and economic institutions at work within them.
81. Though North, Wallis, and Weingast (2013) and Kalyvas (2006) came close.
82. Metz (2017); Mearsheimer and Walt (2007).
83. This image is described succinctly in Almond (1991).
84. As told by Taber (1965).
85. Walt (1996); Waltz (1979).
86. Mearsheimer (2001, 192, 363).
87. Ostovar (2018, 13). Origins of dominant state systems from various disciplines include Barnes (1924); Morgenthau (1948); Lipset (1959); Moore (1967); Huntington (1968); Carneiro (1977); McNeill (1982); Gellner (1988); Tilly (1990); Alon (2005); Scott (2007); Taylor and Botea (2008); North, Wallis, and Weingast (2013); Acemoglu and Robinson (2012a); and Fukuyama (2012).
88. Rousseau (1755). This force refers specifically to the legitimate violence states wield in Weber (1918).
89. Rousseau (1755, 82–3). For further reading on peasant-landlord conflicts, see Moore (1967); Skocpol (1979); Tilly (1990); Scott (1985; 2017); and Fields (2017).
90. Lenin (1916); Luttwak (1968, 10).
91. Öcalan (2017, 39, 44). See Abdullah Öcalan's writings for more on "democratic confederalism." Öcalan authored 13 books mostly dealing with Kurdish autonomy while the sole prisoner at İmralı Island prison in Turkey (Öcalan 2017, xxii). Öcalan helped form the *Partiya Karkerên Kurdistanê* (PKK/Kurdistan Workers Party) insurgency and political party in the early 1980s. See Rabinovich and Valensi (2020) for an example of how Syrian

Kurds have applied these ideas in a time of war.

92. Von Glahn (1957, 291).
93. Kalyvas and Kocher (2009, 337).
94. Kalyvas (2006).
95. Scott (1998, 54).
96. Glahn (1957); Kalyvas (2006); and Scott (1998).
97. Fagan (2000).
98. Thayer (1963). Another group, the Komitadji, were Balkan irregulars active against major powers in the early and mid-twentieth century but with less prominence (Atabaki and Zürcher 2017, 193–4).
99. Arendt (1963, xiii).
100. Morgenthau (1948, 21); cf. Wagner (2007, 49).
101. See Cialdini (1984) and Cialdini and Goldstein (2004) for a useful review of the literature regarding the way compliance and conformity operate within social influence.
102. Acemoglu and Robinson (2012a). In contrast, Moore (1967) centered the same conflicts on social structures.
103. Acemoglu and Robinson (2012a, 372, 398). Huntington (1968) nested the causes of social revolution within "the interaction between political institutions and social forces" (274).
104. Huntington (1968, 274).
105. Scott (1998, 203).
106. Schelling (1960, 32).
107. Quoted in Lenin (1918, 55).
108. Scott (1998, 148–50).
109. Scott (1998, 157).
110. Lenin (1918, 9).
111. Lenin (1918, 28).
112. From the perspective of international relations, conventional wars occur within the horizontal order of anarchy while irregular wars occur in the vertical order of hierarchy (Waltz 1979, 102–16). The anarchic and hierarchic orders are the only two social structures for political Realists. This mirrors some of the concepts of hierarchy in Durkheim (1893; 1912).
113. Joint Publication 1-0 (2017, I-4). Despite this distinction in the doctrine, prominent authors routinely conflate war with warfare, using the terms interchangeably, e.g., Ucko and Marks (2022).
114. Waltzer (1977).
115. Quoted in von der Heydte (1972, 3).
116. Jalali and Grau (2008, 399).
117. Quoted in Taber (1965, 58).
118. Thayer (1963, xvi).
119. West (2008, 4). To overcome that misunderstanding, Thayer (1963) sought to assess causes of irregular wars and their warfare in his important work *Guerrilla*, in which he examined the cases of Kenya, Vietnam, Cyprus, the Philippines, and Cuba to discover "the sources of their amazing power and their hidden weaknesses" (xv). Thayer (1963) concluded his study of irregular war by noting, "The implications for unconventional warfare of the predominance of the political over the military point of view is obvious" and that "practically all the crucial problems of the guerrilla and the counter-guerrilla operations ... are political" (160).
120. Griffith (1937, 32). See, e.g., the book-length argument Peritz and Rosenbach (2012) made in favor of "this energetic national security capability, which did not exist a decade before" (2).
121. Quoted in Mao (1937, 33). Many Western observers, such as Galula (1964), decried using an insurgent's tactics against them.
122. For an example of those in favor, see Galula (1964). An example against is Nagl (2005).
123. Von der Heydte (1972).
124. Gray (2006, 7).
125. Kalyvas and Kocher (2007). Posner differentiated between risk and uncertainty on one hand, and between probability and frequency on the other (in Fukuyama 2007, 7–19), while Mearsheimer (2001) lamented the "rather crude instruments" (411) available to social scientists for defining and testing these types of distinctions.
126. Olson (1959).
127. Rid (2010, 730).
128. Arquilla (2018, 121).
129. Galula (1964, 14).
130. O'Neill, quoted in Taber (1965, viii).
131. Galula (1964, 30–40).
132. Galula (1964, 7).
133. See Taber (1965, 126–32).
134. See von der Heydte (1972, 86–7 notes 3–5) for an early comparative assessment of the operating force sizes in Malaya, Algeria, and Cyprus.
135. Galula (1964, xiv).
136. Bueno de Mesquita, et al. (2003).

137. White (1999, 59–61). Like White (1999), Samuel Griffith identified doctrine, organization, and action as the key elements of a revolutionary movement in his commentary on Mao's (1937) *Yu Chi Chan*.
138. Guevara (1961, 1).
139. See also von der Heydte (1972, 140).
140. Rodriguez (2009, 70, 81).
141. Huntington (1968, 264). See also Skocpol (1979) and Arendt (1963).
142. Gaddis (2018) made just such an argument.
143. Cf. Arendt (1963, 282, note 1).
144. Amenta (2005, 104); Goodwin (2001; 2005, 410).
145. The Arabic name of the group is *ad-Dawlah al-Islāmiyah fī'l-'Irāq wa-ash-Shām*, the Islamic State of Iraq and Greater Syria. Common use in Arabic includes the abbreviation *Da'esh*, which can have a negative connotation. Common English-language abbreviations include ISIS and ISIL. The term "Da'esh" will be used throughout this book for consistency.
146. Kalyvas (2006, 277); Reid (2005, 321).
147. Goodwin (2005).
148. Walt (1996).
149. Arendt (1963, xvi); Skocpol (1979).
150. Huntington (1968, 264).
151. Quoted in Strauss (1987, 304).
152. Huntington (1968, 264).
153. Guevara (1961, 11).
154. Rodriguez (2009, 52). See Acemoglu and Robinson (2012a; 2012b) for an economic history of conditions leading to irregular wars in Latin America from around 1500 to the present.
155. Acemoglu and Robinson (2012a; 2012b). Foa (2016) disagreed with this view.
156. Lenin (1916, 11).
157. Huntington (1968, 264). For examples of differences in understanding revolutions both political and social, see Huntington (1968); Fukuyama (2012); Walt (1996); Arendt (1963); and Skocpol (1979).

Chapter 4

1. E.g., van Creveld, Nagl, Kilcullen, Biddle, and Boot.
2. Kilcullen (2009, xix). Kilcullen took an approach mirroring earlier authors such as Clutterbuck (1980) and Turbiville (2007).
3. Jones and Smith (2010, 261).
4. Gray (2006) rightly noted that "Strategic thought tends to slumber between episodes of security alarm" (3).
5. Van Creveld (1991, 18–21) commits this error among many other errors.
6. Chivvis (2015, 17). Operation Barkhane replaced Operation Serval in 2014, with a focus mainly against violent Salafi-Jihadist groups in and around Mali.
7. Quoted in Osanka (1962, 266). That officer went on to observe how conventional military approaches in these contexts were "cured by the promises of scientists and led into dangerous concepts of warfare, relying too much on machines and forgetting men." Most importantly, he learned that "Absolute technical superiority cannot of itself win cheap wars" (quoted in Osanka 1962, 131, 136).
8. See Osanka (1962, 169).
9. Clausewitz (1832, 112, 120, 121).
10. Jones and Smith (2010) critiqued Kilcullen's (2009) failure to do just that. Pollack (2019) framed well the implications of the missing research when he noted that, "Understanding why these non-state militaries have proven more effective than the state armies we have been trying to improve is more than just an intriguing mystery; it is a vital national interest" (x), though he did not provide a solution to that mystery.
11. E.g., Pollack (2019, 490–1, 502).
12. Pollack (2019, 503). In fact, Pollack mentions this in just a single paragraph over a nearly 600-page book about Arab militaries. He also inaccurately traced the origins of Da'esh, which began from elements forming in the 1980s in Afghanistan and the 1990s in Jordan, describing it instead as spontaneously generating from al-Qaeda in Iraq in 2006. Rabinovich and Valensi (2021) commit the latter error as well.
13. E.g., Biddle (2021). This tactical myopia is clear in his claim that "The more closely one studies the actual behavior of nonstate actors, the less clear the ostensible category distinction with state conventional warmaking becomes" (xvi). Biddle also chose to purposely misspell the word "guerrilla" throughout his book, perhaps in an attempt to delegitimize these actors *vis-à-vis* the state.

Notes—Chapter 4

14. Biddle (2021, xvi).
15. Mumford and Reis (2013, i). And I say "officers" because the subjects are virtually never enlisted, despite the dual burdens of execution and the price of war remaining the common lot of the enlisted man.
16. Mumford and Reis (2012, ii).
17. Clausewitz (1832, 99).
18. Voelz (2014). The concept of scientism has a long history. One prominent example is the biological materialism promoted by figures such as Ludwig Büchner in the nineteenth century (cf. Atabaki and Zürcher 2017, 4).
19. Voelz (2014, 89).
20. Osanka (1962, 100, 275, 278).
21. Öcalan (2017, 44).
22. Fishel and Manwaring (2008, 51).
23. This is Title 10 USC § 127e, "Support of special operations to combat terrorism." Earlier iterations of this initiative existed, most recently § 1207 of the National Defense Authorization Act, but they were not permanently provisioned under Title 10.
24. See Scott (1998) for an analysis of Prussian cameral science in terms of state power.
25. See Fishel and Manwaring (2008) for an example of this cameralist view. See Mao (1937, 71–2) for the seven.
26. Fishel and Manwaring (2008, 40).
27. Osanka (1962, 269).
28. Quoted in Galula (1964, xii) under Nagl's introduction to the revised text.
29. Chivvis (2015, 4).
30. See Smele (2017) for an example of this bounding challenge in a study of the Russian civil wars from 1916–26.
31. Arquilla (2011).
32. Lynn (1996).
33. Pollack (2019, 345).
34. Nagl (2005, 4).
35. Pollack (2019).
36. The best theoretical treatment from a game theoretical perspective of nuclear war theory is still Schelling (1960).
37. Machiavelli (1531); cf. Arendt (1963, 30–2).
38. Boot (2013, 548).
39. Boot followed Liddell Hart (1954) closely in his attitude toward anything that was not a Western or Israeli conventional military force. His database of insurgencies is helpful only as an incomplete list of conflicts from 1770 to 2013, but it is neither all-inclusive nor empirically useful (Boot 2013, 571). Boot failed to format the dates properly in his dataset and he omitted substantively important data. Some important omissions include the war for the Kurdish Mahabad Republic (1946), the Azeri uprising in Iran (1946–47), the Free Papua Movement in New Guinea (1963), the Israel-backed South Lebanon Army as a case distinct from the Lebanese civil war, or Egypt's war in Yemen in 1962. See Waltz (1979), Keohane (1984), and Acemoglu and Robinson (2012b) for examples of more constructive approaches.
40. Boot (2013, 53). See McNeill (1982) for a discussion of the correct "gunpowder empires."
41. Boot (2013, 158). See De Waal (2019) for an accurate history of the Caucasus.
42. Scott (1998).
43. Scott (2017, 17).
44. Tilly (2003).
45. Fukuyama (2007, 2).
46. Trinquier (1964, 15).
47. In Trinquier's case, the status quo meant French colonial domination of Northwest Africa and maintaining subhuman status for non-French Algerians, Tunisians, and Moroccans.
48. Alleg (1958, ix).
49. Tabatabai (2020, 123); Horne (1977).
50. Quoted in Toliver (1997, 47).
51. Quoted in Toliver (1997, 52).
52. These authors include not only French colonial interrogator Roger Trinquier, but also David Galula, John Nagel, David Kilcullen and others. Cornerstone doctrine under this influence includes the U.S. Army Field Manual 3-24, *Counterinsurgency*; the 2014 Joint Publication 3-05, *Special Operations Forces*; the 2017 *Irregular Warfare Joint Operating Concept (JOC) 2.0*; the 2006 *Irregular Warfare JOC 1.0*; the 2007 *Irregular Warfare JOC*, Appendix C; and the DOD's Irregular Warfare Annex to the National Security Strategy (2020).
53. Arquilla (2011, 7).
54. U.S. Army Field Manual 3-24 (2006, xli).
55. CJCS (2019b, II-3).
56. Kuhn (1962, 125).
57. Kalyvas (2006, 32).
58. Kalyvas (2006, 37). Browning reminded observers that "Explaining is not

excusing, understanding is not forgiving" (quoted in Kalyvas 2006, 37).

59. Durkheim (1893).
60. Quoted in Tourison (1991, 84).
61. Jones (2017).
62. Examples include Galula (1964), Trinquier (1964), Kilcullen (2009), and Nagl (2005).
63. Jones (2017, 6). Jones (2017) also observed that "Many practitioners are woefully ignorant of historical trends in past insurgencies and make decisions based on selective cases or intuition rather than sound, objective analysis" (4).
64. Examples include Nagl (2005, 28) and Jones (2017). John Nagl is considered to be "one of the U.S. Army's most-respected thinkers on counterinsurgency" (Rid 2010, 729) and he was a co-author of the U.S. Army Field Manual 3-24, *Counterinsurgency* (2006).
65. Ghaddar (2018, 2).
66. Nagl (2005, 43–4).
67. Gray (2006).
68. Ghaddar (2018, 4).
69. Gray (2006, 43).
70. FM 100-2-2 (1984, 11, 31).
71. NTC 100-91 (1991, 32).
72. FM 34-71 (1982, 10–1).
73. Taber (1965, 6, 10).
74. Taber (1965, 10).
75. Trinquier (1964, 46).
76. Boot (2013, 127); cf. Porch (1982).
77. This example is Nagl (2005, 213).
78. Fishel and Manwaring (2008, 35–8).
79. Chivvis (2015, 6).
80. Chivvis (2015, 33).
81. Chivvis (2015, 17).
82. Perloff (2014, 52).
83. Levy and Thompson (2010, 141).
84. Alden, Thakur, and Arnold (2011, 163).
85. Tilly (2003, 237, 19, 233).
86. Schelling (1960, 122, 160).
87. Schelling (1960, 123).
88. Schmitt (1963, 46).
89. Fearon (2010) lamented that "rhetorical competition leaves people with the impression that each 'ism' must have something to it" (333), while Wagner (2007) saw "a never-ending debate among the competing brands comes to define what the field is" (48).
90. Chekinov and Bogdanov (2013); Beyers (2014).
91. Heras quoted in Ghaddar (2018, 7).

92. For further reading, see "New Generation War" in Bezins (2014) and "cross-domain coercion" in Adamsky (2015). The critical views in Chakrabarti (2010), Evans (2005), and Echevarria (2005) are representative of opponents of the concept. For example, Echevarria (2005) thoroughly critiqued the many weaknesses in the theory of fourth generation warfare, concluding that "We would do well to abandon the theory of [fourth generation war] altogether … it is fundamentally and hopelessly flawed" (vi–2).
93. Lind, et al. (1989).
94. Lind and Thiele (1991); Lind, et al. (1989). See Johnson (2019, 93) for a discussion of Russia's development of war generations.
95. Byely, et al. (1972); Jonsson (2019).
96. Byely, et al. (1972, 250). See O'Hanlon (2019), especially Appendix I, and O'Hanlon (2000).
97. Byely, et al. (1972, 251).
98. Byely, et al. (1972, 251).
99. E.g., van Creveld (1991); Lind and Thiele (1991).
100. This ignorance of the basic elements of war in future conflicts seems to be tied to the continued misunderstanding and misuse of Clausewitz (1832) by proponents of fourth generation war.
101. Lind and Thiele (1991, 41). The Hama massacre in 1983 was part of the Syrian government crackdown on the Muslim Brotherhood, in which the Assad regime indiscriminately murdered up to 40,000 civilians in a single military operation to silence that group's perceived opposition to the regime.
102. Hammes (2006).
103. The studies leading to such conclusions lazily analyze what Acemoglu and Robinson (2012a) identified as "solved political problems" (68) rather than attempting to explain ongoing issues, not unlike many studies in economics.
104. It is epitomized by such authors as Martin van Creveld and Thomas Hammes, and their works are often read in U.S. military training academies.
105. Lynn (1993, 9).
106. Lynn (1993, 9). Lynn did go on to note that at last six major flaws exist in van Creveld's writing on military supply and logistics, while arguing van Creveld was "flat wrong" (107) in his interpretation of

historical sources. Still, both Lynn's glancing criticism and van Creveld's influential neoconservative *Weltanschauung* remain firmly within conventional war theory. It is unsurprising that many proponents of fourth generation warfare were also neoconservatives who sought to overthrow states in the Middle East after 2001 at Israel's prompting.

107. Lind and Thiele (1991, 121).
108. Chakrabarrti (2010, 73).
109. See Jonsson (2019, 107) for an argument that the perception of the nature of war can change in an evolutionary way. See also his critique of the revolution in military affairs, network-centric warfare, and hybrid warfare. For a summary of critiques of the "gray zone" concept, see Jonsson (2019, 16). See Shurkin, Cohen, and Chan (2022) for a brief overview of network-centric warfare along with an analysis of French applications of the concept. For a proponent's view of lawfare, see Goldenziel (2020). For a Chinese perspective of the American mosaic warfare and kill web concepts, see Qi, et al. (2021).
110. Gray (2006).
111. Gray (2006, 51).
112. Schnaufer (2017, 23).
113. Stoker and Whiteside (2020, 30); Jonsson (2019, 9).
114. Lambakis (2004, 13, 39).
115. Stoker and Whiteside (2020).
116. Gray (2002); cf. Bickel (2004).
117. Biddle (2021, 301).
118. Consider, for example, how Cordesman (2017) employed the term in calling for a "revolution in civil-military affairs," essentially making this a euphemism for development in any realm of military activity.
119. As Kalyvas (2006) noted, "The wrong research path would be to again coin conceptual categories grounded in current events rather than good theory" (117).
120. Clausewitz (1832, 17).
121. See Waltz (1959), Rosecrance and Stein (1993), and Mearsheimer (2011) for compelling arguments that war is the result of foreign policy failure. Joint Publication 1-0, *Doctrine of the United States Military*, also acknowledges that war is a "result from failure of states to resolve their disputes by diplomatic means" (I-2).

Chapter 5

1. Kissinger (1969), quoted in Nagl (2005, 171).
2. See Mearsheimer (2001, 385–6) for further discussion of conventional war strategies.
3. Mearsheimer (2001, 246); Cleveland and Egel (2020).
4. Title 50 USC, Ch. 3, § 21; Joint Publication 1-0 (2017, I-2).
5. Waltz (1979, 103).
6. Title 50 USC, Ch. 33, §1541.
7. Title 50 USC, Ch. 33. See also § 1208 of the 2006 U.S. National Defense Authorization Act, superseded by Title 10 USC § 127e. And cf. § 1207 of the 2005 U.S. National Defense Authorization Act *contra* Title 10 USC, Ch. 16, § 333 for more perspectives on U.S. defense funding for building partner capacity.
8. Joint Publication 1-0 (2017, I-5).
9. Public Law 107-40.
10. Gray (2006, 43).
11. Joint Publication 1-0 (2017, I-3).
12. Joint Publication 1-0 (2017, I-2), emphasis added.
13. Joint Publication 1-0 (2017, I-3).
14. Mearsheimer and Walt (2007, 63).
15. Jones (2017, 6). This intellectual tradition was passed via Galula, Trinquier, Taber, and Callwell, among others.
16. Sun Tzu, 3.18.
17. Joint Publication 1-0 (2017, I-4).
18. For example, see the misinterpretation Tabatabai (2020, 205) supported at length in her analysis of Iran's decision to remain in the Iran-Iraq War after 1982.
19. Machiavelli (1531, 37).
20. See Bassiouni (2008) for a detailed legal analysis of the problems with this approach and its tendency toward disproportionate non-combatant deaths in wars between states and non-state actors. See Cleveland and Egel (2020) for a detailed study of a segment of U.S. irregular warfare failures.
21. It is not difficult to imagine Israeli military and political leaders defending themselves at a court similar to the International Criminal Tribunal for the Former Yugoslavia if Hezbollah or Hamas had the same level of support at the U.N. as Israel. This is especially true during their wars against Lebanon (1968–2006) and their wars against Gaza. In those wars, Israeli

forces killed tens of thousands of civilians using numerous methods barred by international law and the U.N. charter, including dropping cluster bombs on densely populated urban centers, collective punishment, airstrikes on non-military targets, and so on.

22. Gray (2006, 3, 9–10). Those five categories are foreign internal defense, stability operations, unconventional warfare, counterterrorism, and counterinsurgency (Joint Publication 3-05 2014).
23. Joint Publication 3-05 (2014).
24. Department of State (2006). This allowed DOD to transfer the money to DOS so that, in part, the activity could be programmed as Security Assistance money rather than Security Cooperation money. Recipients included the Lebanese Internal Security Force, and money also went to support programs in Yemen, Indonesia, Mali, Niger, Mauritania, Malaysia, Philippines, Ethiopia, Kenya, Nepal, Haiti, and Colombia. Section 1207 of the 2006 NDAA became statute in 2008 when it was written into § 127e of Title 10 USC.
25. Joint Publication 3-20 (2017).
26. Eisenstadt and Pollack (2020, 97).
27. Joint Publication 1-0 (2017, I-6). "Anti-Westphalian" in this sense means that the war does not comport with the traditional military activities of the conventional forces of a state built around the Westphalian model of state sovereignty.
28. CJCS (2019a, v).
29. Culprits include Biddle (2021) and van Creveld (1991).
30. Biddle (2021, 4); CJCS (2019a).
31. The specific Realists this approach follows include Waltz (1959; 1979); Walt (1996); and Mearsheimer (2011). The Realists who identify that their theories lack certain elements include Keohane (1984); Fearon (1995); and Mearsheimer (1994).
32. Jasper (2005, 125).
33. Quoted in Mansfield (1987, 689).
34. E.g., Hooks and Rice (2005, 571).
35. Amenta (2005, 96–119).

Chapter 6

1. From a 1946 interview, quoted in Meiselas (1997, 185).
2. Othen (2015, 98).
3. Nietzsche (1886).
4. Jomini (1862).
5. Jomini (1862, 102).
6. Jomini (1862, 215).
7. McNeill (1982) described the conventional forces of the state as "An artificial community bureaucratically structured and controlled ... based on deep-seated, stable, and very powerful human sentiments. What an instrument in the hands of statesmen, diplomats, and kings!" (133).
8. Huntington (1957, 2).
9. Tilly (1990, 170–81); Wagner (2007, 25); and Fearon (2010, 336).
10. Ostovar (2018, 13).
11. For further reading, see Herz (1950); Waltz (1959); Jervis (1978); Waltz (1979); Fearon (1995); Mearsheimer (2001); and Wagner (2007).
12. Wagner (2007, 26); cf. Fearon (2010, 341).
13. Gorka and Kilcullen (2011, 15).
14. See Showalter (2004, 118–34) for a summary of the development of the Prussian military state.
15. Huntington (1957, 30); Showalter (2004, 118).
16. Schmitt (1963, 23).
17. Quoted in Showalter (2004).
18. Huntington (1957, 197); Voelz (2014, 85).
19. Voelz (2014, 85).
20. Huntington (1957, 263). The causes of this transformation are beyond the scope of this book. Some studies of these causes can be found in Huntington (1957); Waltz (1959); and Murray, Knox, and Bernstein (1996).
21. Clark (2002, 119). Clark described the American way of war in terms of its ability to "muster an overwhelmingly large force; prepare and train it; then use it to achieve militarily decisive results" (450).
22. Gray (2006, 30).
23. Gray (2006).
24. Jomini (1862, 212).
25. Jomini (1862, 212).
26. Mahan (1890, 55); Huntington (1957).
27. Melson (2019, 4).
28. Pollack (2019, 53); Clausewitz (1832).
29. Clausewitz (1832, 544).
30. Sun Tzu, 23.
31. Mao Zedong (1937, 51).
32. Luttwak (1987, 132).
33. Stülpnagel (1924).
34. McNeill (1982, 130).

Notes—Chapter 6

35. McNeill (1982, 130–4, 140).
36. For further reading, see Luttwak (1976); Hildinger (2002); and Boatwright, Gargola, and Talbert (2004).
37. McNeill (1982, 140).
38. Atabaki and Zürcher (2017, 209–10).
39. Luttwak (1987, 91).
40. Fravel (2019, 22).
41. Fravel (2019, 64).
42. Fravel (2019, 40).
43. Quoted in Reid (2006, 210).
44. Reid (2006, 210).
45. See Lambert (1990) in defense of the Crimean War being the last and see Reid (2006) in defense of it being the first modern war.
46. Clausewitz (1832, 188).
47. Clausewitz (1832, 391).
48. Luttwak (1987, 132).
49. Clausewitz (1832, 639).
50. Emmerich (1789).
51. Emmerich (1789).
52. Simes (1767) defined the partisan as "a person very dexterous in commanding a party; and who, knowing the country well, is employed in getting intelligence or surprising the enemy's convoy" (5). Simes authored several influential works on British and American military matters, read widely from the Peninsular Wars (1808–14) to the Spanish-American War (1898) (Kehoe 2000, 214). Some of Simes's most influential works include *The Regulator* (1780), *A Military Course* (1777), *The Military Guide for Young Officers* (1772), and *The Military Medley* (1767). Around the same time, Prussian Baron General von Steuben wrote the U.S. Army's infantry manual, known unofficially as "the Blue Book." Von Steuben's manual was used as official U.S. Army infantry doctrine from 1779 until 1812.
53. Joint Publication 3-05 (2017).
54. Taber (1965, 2); Chandler (1989).
55. Chandler (1989, 8).
56. Chandler (1989, 10).
57. Cf. MCIP 3-11.01 *USMC Combat Hunter Program*.
58. Thayer (1963, 128).
59. Thayer (1963, 128).
60. Huntington (1957, 64).
61. Melson (2019, 12).
62. Melson (2019, 99).
63. Melson (2019, 6). See Birtle (2007, 131–57) for a historical narrative of how U.S. military doctrine borrowed German irregular warfare concepts wholesale after the Second World War.
64. Melson (2019, vi).
65. Birtle (2007, 133).
66. According to Birtle (2007), these writings "would exert a profound influence over American doctrine for many years" (133).
67. Quoted in Melson (2019, 165).
68. Melson (2019, 162).
69. Melson (2019, 163).
70. U.S. Army Field Manual 3-24 (2006).
71. Boot (2013, 541).
72. Sewell, quoted in Field Manual 3-24 (2006, xxxv, xxxix); Rid (2010).
73. Specifically, T.E. Lawrence (1926); Galula (1964); and Trinquier (1964).
74. Castelli (2009, 18).
75. Horne (1977, 102).
76. Horne (1977, 254).
77. Horne (1977, 26).
78. See Sloan (2019, 127–8) for a discussion of some theories of organizational learning in a military context.
79. Caverley (2010, 142). Nagl (2005) is an example of a "military myopic" treatise.
80. See Bayley and Perito (2010, 69) for an important discussion of the role of police in war, especially during insurgencies and stability operations.
81. Luttwak (1987, 134).
82. Fritz, quoted in Melson (2019, ix).
83. Fearon and Laitin (2003, 88), emphasis in original.
84. Staniland (2015, 773).
85. Staniland (2015, 773–6).
86. E.g., Army Field Manual 3-24 (2006); Henriksen (2007); Turbiville (2007; 2009); Celeski (2010); and van Creveld (1991).
87. Joint Publication 3-05.130 (2008).
88. Staniland (2015, 776).
89. See Rid (2010) for an excellent overview of this historical development through his study of David Galula and other French colonial officers.
90. Jomini (1862, 22). See also Chandler (1989) for a fuller account of this campaign.
91. Porch (1982, 275).
92. Callwell (1896, 99).
93. Porch (1982, 294).
94. Miller (1977, 546).
95. Barr (2011, 137).
96. Taber (1965, 64). See Turbiville (2007) for a recent celebration of this failed French strategy.

97. Horne (1977, 111).
98. Galula (1964, 69). Schmitt (1963, 47) measured 200,000 French soldiers against 30,000 Algerian partisans, a ratio of almost 7 to 1. Trinquier (1964, 7), a French *para* who participated in the war put French numerical superiority to the FLN at 10 to 1.
99. Horne (1977, 198).
100. Horne (1977, 167), emphasis in original.
101. Trinquier (1964); Broadwell (2012); McChrystal (2013); Westmoreland (1976).
102. Horne (1977, 259).
103. Horne (1977, 190).
104. Horne (1977, 96, 207). This line of thinking directly influenced misguided writers on the subject such as Pollack, Kilcullen, and Boot.
105. Horne (1977, 99).
106. Quoted in Horne (1977, 307).
107. Osanka (1962, 437).
108. Horne (1977, 334).
109. Pollack (2019, 54).
110. Azizi (2020, 130–1).
111. Osanka (1962, 259).
112. Chivvis (2015, 12).
113. Chivvis (2015, 13).
114. Chivvis (2015, 131–8).
115. See, e.g., Liddell Hart (1950) and Tuker (1948).
116. Porch (1982, 19). Britain's use of automatic weapons provided a decisive edge.
117. Taber (1965, 100–1).
118. Osanka (1962, 436). The German Empire also used concentration camps in its own colonies in Namibia shortly after Britain concluded its Boer Wars in nearby South Africa. Spain, Italy, and the U.S. all used concentration camps in their own colonial policing conflicts (Forth and Kreienbaum 2016).
119. Allison (2001, 102).
120. Allison (2001, 127). Innovations in colonial policing activities like the use of air power were not exclusive to Britain. Aircraft were also first used to bomb ground-based targets in Italy's strikes against irregulars in Libya after Italian forces seized Libyan territory for colonial expansion shortly after the invention of the airplane (Carson 2018, 90). One of the earliest dive-bombings in the history of air warfare occurred midway through the U.S. Marine Corps occupation of Nicaragua (1912–33) (Cabezas 1982, 224).

121. Kemp (1959, 18).
122. Quoted in Barr (2011, 43).
123. Figures given in Taber (1965, 131–2).
124. Von der Heydte (1972, 87).
125. Taber (1965, 138–41); Von der Heydte (1972, 86–7, notes 3–5).
126. Nagl (2005) being a leading supporter of British conduct in that war.
127. Taber (1965, 141).
128. Corum (2006). This was largely because the insurgents in Malaya were foreigners and did not represent Malayan interests, while the insurgents in Cyprus were Cypriots representing Cypriot interests.
129. Kalyvas (2006, 90).
130. See Striffler and Moberg (2003); De Soto (2001); and Rodriguez (2009) for perspectives on the effects of U.S. involvement in Latin American states.
131. U.S. Marine Corps (1940, 1).
132. See Butler (1935) for a first-hand account of U.S. Marine Corps involvement in these conflicts.
133. Thayer (1963, 44).
134. Nagl (2005, 122). See U.S. Army Field Manual 31-21 (1958/1961) *Guerrilla Warfare and Special Forces Operations* for the doctrine in place when America became involved in Vietnam. See Turbiville (2007) for an example of the persistence of these same calcified views among American military theorists. Cleveland and Egel (2020) argued that U.S. Special Operations Forces in the Middle East in the 2010s still operated on the near-legendary assumptions "entombed in such seminal documents as the counterinsurgency field manual or *Small Wars Manual*" (150).
135. Kalyvas (2006, 90).
136. Giap (1970, 12).
137. According to U.S. Army Field Manual 6-20, chapter 2, "Unobserved fire is fire for which the points of impact or burst are not observed. It involves predicting where targets are, or will be, and placing fire on them. Use of unobserved fire requires follow-up activity to assess effectiveness." Hawkins (2013) conducted a detailed study of this issue and found that "Excessive unobserved firepower expenditures by Allied forces during the Vietnam War defied the traditional counterinsurgency principle that population protection should be valued more than destruction of the enemy" (ii).

Notes—Chapter 6

138. Taber (1965, 85).
139. Taber (1965, 85, 93).
140. Taber (1965, 92–3).
141. Giap (1970, 12).
142. Quoted in Gray (2006, 37).
143. Quoted in Chehab (2005, 10).
144. Biddle (2021, 55).
145. Mearsheimer (2001, 96).
146. Simatupang (1972, 76).
147. Quoted in Tourison (1991, 276).
148. Chehab (2005, 30).
149. Chehab (2005, 25).
150. Lukens (2010, 7). Gray (2002) argued that the U.S. "traditionally has been all but comprehensively uninterested in irregular warfare" (2).
151. Kilcullen (2012, 20).
152. Stern and McBride (2013, 5).
153. Gaddis (2002).
154. In a precise example of this entanglement, Gray (2005) blurred together "terrorism and other forms of irregular warfare" (2) in his study of the latter.
155. Bayley and Perito (2010, 72).
156. Baker, et al. (2006, 94).
157. Garrett-Peltier (2019).
158. Crawford (2019).
159. Biden, quoted in Helman and Tucker (2021).
160. Vine, et al. (2020).
161. Vine, et al. (2020, 3).
162. Vine, et al. (2020, 24).
163. Savell (2019).
164. Savell (2019).
165. Martowych (1950, 4).
166. Quoted in Rogers (2016, 31).
167. Mearsheimer (2001, 109).
168. CENTO (1955).
169. Caverley (2010).
170. Bill (1988, 147).
171. Taber (1965, 78).
172. Kalyvas (2006).
173. Rodriguez (2009, 120–32).
174. Quoted in Nagl (2005, 128).
175. Biddle (2021, 118).
176. Rajagopalan (2007, 84).
177. Chakrabarti (2010, 73).
178. Indian Armed Forces (2006); Dixit (2010, 133).
179. Tilly (2003, 173).
180. An example of this phenomenon in an irregular war occurred in the first half of 2016. Between Mosul and Erbil, U.S. and Kurdish forces established a FLOT a few kilometers away from the Da'esh FLET. Between this zone, relatively little activity occurred as both sides dug in and displayed their combat capability to the other with relatively little action in the dead space, save for the occasional suicide run of Da'esh fighters or explosive-laden "zombie" trucks. The main combat action during this time took place instead deep inside Da'esh-controlled territory surrounding Mosul.
181. Walt (1996, 152).
182. Walt (1996, 2).
183. Waltz (1979, 138).
184. Quoted in Walt (1996, 149).
185. Smele (2017); De Waal (2019).
186. Data collected in Walt (1996, 150); cf. Carson (2018). Compare this to the equipment Britain gifted the Hessian partisans who fought against the colonies in the American Revolution (Emmerich 1789).
187. Schmitt (1932, 53).
188. Von der Heydte (1972, 181).
189. Robespierre (1792).
190. Walt (1996, 150).
191. Walt (1996, 150).
192. Carson (2018); Kemp (1957); and Orwell (1938).
193. Kemp (1957, 28).
194. Churchill quoted in Osanka (1962, 330). Data on number of divisions is from von der Heydte (1972) and Luttwak (1987).
195. This theory holds that "whether state goals are conservative or assertive, status quo or revisionist, major powers have powerful and overlapping reasons to view large-scale escalation … as counterproductive for geostrategic, economic, or domestic political reasons" (Carson 2018, 46).
196. Carson (2018, 27).
197. Carson (2018, 6, 29).
198. Tilly (2003, 200).
199. Salibi (1988, 125, 142).
200. Mahan (1890, 88).
201. Mahan (1890, 107).
202. Mahan (1890, 133).
203. Mahan (1890, 205).
204. Kemp (1957, 12–33). See Kemp (1957, 24–5) for a list of men, weapons, and equipment Germany and Italy provided to the Nationalists and the support Russia gave the Republicans.
205. Shinoda (2018, 28).
206. Waltz (1979) argued that "If a country, because of internal disorder and lack of coherence, is unable to rule itself,

no body of foreigners, whatever the military force at its command, can reasonably hope to do so. If insurrection is the problem, then it can hardly be hoped that an alien army will be able to pacify a country that is unable to govern itself" (188).
207. Goodwin (2001, 123).
208. Taber (1965, 88).
209. Traboulsi (2012, 137). See Birtle (2007, 183–95) for a summary of this operation and its direct impact on the formation of irregular warfare doctrine in the U.S. military.
210. Chehabi (2006, 166).
211. Michaels (2011); Thompkins (2012, 525). See UN S/15194, dated 10 June 1982 for detailed figures of Israeli aggression in Lebanon from 1978 to 1982.
212. Jabbra and Jabbra (1983).
213. The original name was Islamic Jihad and then changed to Hezbollah a few years later (Ostovar 2018, 112, 146).
214. Jabbra and Jabbra (1983); Barr (2018); Little (2004); and Ostovar (2018). The Free Lebanon State was an unrecognized state that Israel established in 1979 comprising a strip of land in South Lebanon on the Lebanon-Israel border. The Free Lebanon State ended in 1984.
215. Ostovar (2018, 116–7).
216. Ostovar (2018, 6).
217. Clark (2002, 55).
218. Luttwak (1987, 258).
219. Hovsepian (2008); Waltzer (1977).
220. Bennett (2013, 76).
221. Bennett (2013).
222. Both quotes from Bennett (2013, 92).
223. Quoted in Sherman (2009, 19).
224. Sherman (2009, 19).
225. Vine, et al. (2020, 13).
226. Malicdem (2017).
227. Savage and Caverley (2017).
228. Savage and Caverley (2017, 554).
229. Eisenstadt and Pollack (2020, 97).
230. El-Sisi (2006).
231. Robinson (2010, 22).
232. Robinson (2010) astutely pointed out that "Rather like their counterparts in the U.S. Army after Vietnam, Russian officers seem to have decided that the main lesson to draw from counterinsurgency experience was not to engage in counterinsurgencies. Instead, they chose to refocus on studying conventional inter-state warfare" (22).
233. Fearon and Laitin (2003).
234. Luttwak (1987, 170).
235. Similarly, Liddell Hart (1954) argued that "The development of nuclear weapons would tend to nullify their deterrent effect, thereby leading to the increasing use of guerrilla-type strategy" (xv).
236. Hammond (2015, 316).
237. Both quotes from Suhrke and De Lauri (2019, 14).
238. Suhrke (2008); Suhrke and De Lauri (2019).
239. See Bew (2016) for a detailed discussion of the nuances inherent to the concept of *Realpolitik*.
240. Galeotti (2019, 14).
241. USASOC (2015, 15).
242. Quoted in Jonsson (2019, 81).
243. USASOC (2015, 14).
244. Alden, Thakur, and Arnold (2011, 6).
245. Jentzsch, Kalyvas, and Schubiger (2015, 755).
246. Aliyev (2016, 500).
247. Bou-Nacklie (1993, 648).
248. LeRiche and Arnold (2012, 94).
249. North, Wallis, and Weingast (2013, 168).
250. See, e.g., Barr (2011); Bill (1988); Keohane (1984, 64); Rodriguez (2009); Van Goor (2004); and Bickerton and Klausner (2015).
251. van Goor (2004, 7).
252. Welker (2014).
253. Acemoglu and Robinson (2012b, 340).
254. Devlin (2007); Trinquier (1964); and Othen (2015).
255. Othen (2018, 11).
256. Othen (2018, 132–8).
257. Rodriguez (2009, 13).
258. Rodriguez (2009, 18).
259. Rodriguez (2009, 50). See Mearsheimer (2001, 234–66) for a discussion of Manifest Destiny, the Monroe Doctrine, and offshore balancing as a grand strategy.
260. Chivvis (2014, 16).

Chapter 7

1. Wilson (1918).
2. The 1215 baronial revolt was not unlike the *mashruteh* (Constitutional Revolution) in Iran in the summer of 1906 or the Tennis Court Oath on 20 June 1789

113. Quraishi (2004, 30).
114. Acemoglu and Robinson (2012a, 7).
115. Rodriguez (2009, 52).
116. Jones and Rideout (2015, 136–41).
117. Knetsch (2003, 59).
118. Quoted in Knetsch (2003, 58).
119. Knetsch (2003, 49–50).
120. Kropotkin (1892).
121. Quoted in Chivvis (2015, 55).
122. Quoted in Chehabi (2006, 241). Indeed, Vine, et al. (2020) found that "Political instability and violent conflict have heightened and been mutually reinforcing with humanitarian crises caused by drought, flooding, attendant famine, and widespread poverty" (12).
123. Quoted in Traboulsi (2012, 185).
124. Taber (1965, 101).
125. Guevara (1961, 120).
126. Quoted in Horne (1977, 95). The *Front de libération nationale* (National Liberation Front/FLN) was the political wing of the Algerian insurgency against France.
127. Guevara (1961, 10).
128. Salibi (1988, 112).
129. Allison (2001).
130. Allison (2001, 127).
131. Martowych (1950, 9); Chehabi (2006).
132. Kalyvas (2006, 258).
133. This political arm was the *Ethnikó Apeleftherotikó Métopo* (National Liberation Front); the coercive arm was the *llinikós Laïkós Apeleftherotikós Stratós* (Greek People's Liberation Army).
134. Quoted in Stalin (1924, 37).
135. Stalin (1924, 37).
136. Solahudin (2013, 35).
137. Taber (1965, 85).
138. Taber (1965, 32, 86).
139. Taber (1965, 76, 92).
140. Che Guevara (1961) considered this to be "the head of a large movement with the characteristics of a small government" (78, 82).
141. Guevara (1961, 86).
142. Guevara (1961, 88).
143. Simatupang (1972, 70).
144. Simatupang (1972, 95).
145. Simatupang (1972, 70).

Chapter 8

1. Quoted in Guevara (1961, 88).
2. Luttwak (1968, 147–69).
3. Simatupang (1972, 71).
4. These compound motivations are based, in part, on Margaret Levi's motivation framework found in Kalyvas (2006, 101).
5. Third-party intervention in support of the insurgency is unconventional warfare; in support of the state, it is counterinsurgency.
6. In the case of the U.S., this usually involves a combination of authorities drawn from Titles 10, 18, 22, and 50 of the U.S. Code, among other authorities and statutes.
7. Forsythe (2010, 57).
8. Hobbes (1651, 721–28).
9. See also Milgram (1963), Cialdini (1984), Zimbardo (2007), and Forgas and Williams (2001).
10. Helvey (2004, 23).
11. Öcalan (2017, 76).
12. Forsythe (2010, 223). The other five sources are reward, coercion, referential, expert, and information. See also French and Raven (1959).
13. Smele (2017, 89).
14. Bueno de Mesquita, et al. (2003, 370).
15. See Acemoglu and Robinson (2012a, 123; 2012b) for a political-economic discussion of this issue and see Geddes (2009) for a rational choice perspective.
16. Kalyvas (2006, 46).
17. Arendt (1963, 68).
18. Von der Heydte (1972, 124); Waltz (1959, 149).
19. Barr (2011, 138). McNeill (1982) similarly observed that "An enemy at the gates has always been the best substitute for spontaneous consensus at home" (380) and "an evident outside threat was … the most powerful social cement known to mankind" (382).
20. Acemoglu and Robinson (2012a, 21); von der Heydte (1972, 45).
21. See Speier (1941, 445), Barber (1951), and Fukuyama (1992, 162, 183) for a discussion of *gloire* and see Fukuyama (1992) for a discussion of Plato's *thymos*. Other unique drivers include Islamic *tasbih* (glorifying Allah) and Roman *virtù* (virtue).
22. Chehab (2005, 142).
23. Levi, quoted in Kalyvas (2006, 101).
24. Quoted in Horne (1977, 214).
25. Quoted in Horne (1977, 483).
26. Quoted in Chehab (2007, 98).

27. Kalyvas (2006, 92). See Perloff (2014, 141) for a summary of the research on attitudes and behavior.

28. Based in part on the motivation construct Margaret Levi proposed (in Kalyvas 2006, 101).

29. Bechara (2003, 140).

30. Chehab (2007, 23).

31. Quoted in Acemoglu and Robinson (2012b, 373–5).

32. The RUF failed as an insurgency but succeeded later as a political party.

33. Loadenthal (2016).

34. LeRiche and Arnold (2012, 26).

35. LeRiche and Arnold (2012, 82).

36. Chehab (2007, 196).

37. Guevara (1961, xv). Bill (1988) provided an illustration of this claim in his description of the situation in Iran in the 1930s and 1940s: "Far more intolerable to Iran than the financial consideration [of oil revenues] was the question of national sovereignty. The institution of a concession by which large foreign corporations had complete control of resource exploration and exploitation over large areas of land for long periods of time galled the nationalists" (63).

38. Quoted in Avrich (1976, 157).

39. Martowych (1950, 7).

40. Thayer (1963, 70).

41. Horne (1977, 39).

42. Kemp (1959, 81).

43. Chehab (2007, 16–23).

44. The "accidental guerrilla syndrome" was popularized by the influential military commentator David Kilcullen, especially through his 2009 book on the subject. See Giap (1970) for a collection of Giap's own interviews and writings at the height of the U.S. involvement in the Vietnam war.

45. Giap (1970, 15).

46. Irwin (2019, 109); Reynolds (2005).

47. Giap (1970, 20).

48. Despite the appeals to pathos and ethos in Israel's *hasbara*, Israel's invasion of Lebanon in 1982 featured economic interests not unlike the American actions in Guatemala up to 1955 (Shahak 1997, 91). While the independent United Fruit Company was the economic catalyst for the coup in Guatemala, it was the state-owned agricultural monopoly Agrexco that benefited from Israel's invasion by stifling Lebanon's own trade in produce with its Arab neighbors, especially Syria (Shahak 1997, 102–22). For example, on Wednesday, April 4, 1984, the Hebrew-language edition of the Israeli newspaper Haaretz reported that Israel was exporting a third of its fresh produce directly to Lebanon via the Agrexco monopoly, while the rest went primarily to Europe (Schechla 1984). A significant element of the difficulty Israel faced in Lebanon with its proxy Christian South Lebanon Army and puppet government in Beirut revolved around customs, duties, commodity movement, and licenses for non-Jewish merchants to sell in Israel rather than existential justifications for war that *hasbara* portrayed. That article and hundreds more never made it out of the Hebrew-language domestic media due to Israeli government censorship and control of the press. For more on early Agrexco, see Doerr, Coling, and Kerr (1970) and Shehadeh (1988).

49. See Dingel (2013, 70) and Ostovar (2018) for more detailed information on how Israel's invasion of Lebanon directly instigated Hezbollah's creation and see Mallison (1983) for a contemporary account of events. See Nerguizian (2017) for an overview of Hezbollah's superior military efficacy *vis-à-vis* the Lebanese Armed Forces.

50. Hovsepian (2008, 65). Hezbollah did not always adhere to this decision, with elements under Iranian and Syrian control within Hezbollah conducting various assassinations, subversive activities, and sabotage against political opponents of those two states inside and outside of Lebanon over the years following the Taif Agreement.

51. Hovsepian (2008, 81).

52. Hovsepian (2008, 265); Dingel (2013, 71).

53. Osanka (1962, 330–1).

54. Quoted in Von Dach (1957, iv).

55. Rothchild (2014, 68).

56. Hari (2008, 120).

57. Mearsheimer and Walt (2007, 340).

58. Hovsepian (2008, 5).

59. Hovsepian (2008, 173).

60. Biddle (2021, 110).

61. Hovsepian (2008, 82).

62. Quoted in Hovsepian (2008, 177).

63. Azizi (2020, 202).

64. Azizi (2020, 202). Israel used a similar strategy in Gaza especially with Oper-

ation Cast Lead (2008–09), which killed nearly 1,500 Gazan civilians. The Israeli strategy of collective punishment through indiscriminate violence may have produced limited, short-term successes in the occupied West Bank and Gaza, but it also helped Hamas solidify its position both in military recruitment and political power, just as Hezbollah did.

65. Goodwin (2001, 234–5).
66. Allison (2001).
67. Quoted in Allison (2001, 129).
68. See Loqman (2021) for an excellent summary of events leading to Kurdish statelessness.
69. Meiselas (1997); Knapp, et al. (2016, 21).
70. The authoritative documentary history of the Kurds remains Meiselas (1997), while Eagleton (1963) presented the most comprehensive history of the 1946 war for the Mahabad Republic. One could argue that Self-Administered Northeast Syria approached a similar apogee. Still, statehood continues eluding the Kurds.
71. Eagleton (1963, 27).
72. Eagleton (1963, 27).
73. Eagleton (1963, 49–51).
74. Eagleton (1963, 26, 34, 39).
75. Eagleton (1963, 62, 86); Meiselas (1997, 176).
76. Eagleton (1963, 30–2).
77. Eagleton (1963, 30).
78. Eagleton (1963, 78).
79. LeRiche and Arnold (2012, 27); Meiselas (1997, 308–45).
80. Aliyev (2016, 503); LeRiche and Arnold (2012, 73–78). Not to be confused with Exercise Bright Star, the bilateral Egypt-U.S. event held every two years since the signing of the Camp David Accords in 1978.
81. LeRiche and Arnold (2012, 2).
82. Jones (2017, 3). Many examples of this arrangement exist, such as in Algeria, Indonesia, Sudan, Palestine, Lebanon, Colombia, China, and so on.
83. Clark (2002, 406).
84. Öcalan (2017, xviii).
85. Tilly (2003, 39).
86. Svolik (2012, 168).
87. Kalyvas (2006, 106–10).
88. Taber (1965, 95).
89. Galula (1964, 12).
90. Jones (2017, 38).
91. Jones (2017, 89).
92. Martin and Nordstrom (1992, 202).
93. Knapp, Flach, and Ayboga (2016); Yesiltas and Ozcelik (2018, 24).
94. Hovsepian (2008, 64); Hackett (2018, 20).
95. Taber (1965, 23). These are sometimes to referred to as auxiliary, underground, base, and support as in the U.S. Joint doctrine for unconventional warfare (Joint Publication 3-05.1 2015).
96. Martowych (1950, 44).
97. Byman (2007, 15); Osanka (1962, 181).
98. Alden, Thakur, and Arnold (2011, 153–9); cf. Byman (2007).
99. Quoted in Chehabi (2006, 254).
100. Horne (1977); Trinquier (1964, 9–11).
101. Ostovar (2018, 41).
102. Ismail (2018, 38); Rabinovich and Valensi (2021).
103. Solahudin (2013, 35); Simatupang (1972, 100).
104. Pittaway (2004, 40–51).
105. See Luttwak (1968) for the media's role in a *coup d'état*, something Shariati (1979, 9) described as revolution without the people.
106. Quoted in Acemoglu and Robinson (2012b, 462).
107. Cialdini and Goldstein (2004, 594).
108. Bernays (1928) noted that "In the field of propagandism there is hardly a more powerful method of arousing and controlling public opinion" (33) than propaganda on visual media.
109. Quoted in the introduction to Bernays (1928, 27).
110. Quoted in Barr (2011, 271).
111. Quoted in Horne 1977, 190).
112. Huntington (1957, 90).
113. Walt (1996, 25).
114. *Attitude* in this context is a "learned, global evaluation of an object (person, place, or issue) that influences thought and action" (Perloff 2014, 88).
115. Perloff (2014, 82).
116. Perloff (2014, 141).
117. See Thompkins (2015, 37–8) for several more examples of the use of prisons as recruitment venues for insurgencies and underground movements during war.
118. Quoted in Baniyaghoob (2013, 105).
119. Horne (1977, 111).
120. Quoted in Horne (1977, 71).
121. Ostovar (2018, 51). This process of

incarceration leading to a radiation of ideas that actually benefit the incarcerated individual is not limited to irregular wars. Hitler wrote *Mein Kampf* (1925) while in prison. Karl Dönitz developed the *Rudeltaktik* (wolfpack) strategy for employing U-boats while a prisoner of war in a British camp during the First World War.

122. Chehab (2005, 120). See the unclassified executive summary of the 2014 Senate Select Committee on Intelligence report on detainee abuse for a detailed look at some of these activities and their effects.

123. Quoted in Chehab (2005, 116).

124. Examples include Guevara (1961); Marighella (1969); von Dach (1957); Simatupang (1972); Serge (1951); Azzam (1988); Trotsky (1930); Lenin (1906); Orwell (1938); Hemingway (1940); Martowych (1950); Mao (1937); Lawrence (1920; 1926); Kemp (1957; 1958; 1961); H. (1956); Giap (1970); Gandhi (1961); Fanon (1961); Emmerich (1789); Öcalan (2017); and Cabezas (1982).

125. Kemp (1957); Orwell (1938); and Hemingway (1940).

126. Von Dach (1957); Marighella (1969).

127. Ehrhardt (1935); Von der Heydte (1972); and *Bandenbekämpfung* (1944).

128. Kinzer (1985); Cabezas (1982).

129. Nordstrom and Martin (1992, 204); Sluka (2000).

130. Azzam (1988).

131. Lukens (2010, 6). Al-Qaeda specifically adopted these three volumes: Clausewitz (1832); Mao (1937); and Taber (1965). Da'esh later exploited the old al-Qaeda "join the caravan" call to pull an unprecedented number of foreign fighters into its insurgency after 2014.

132. Simatupang (1972, 155).

133. Bernays (1928, 9).

134. Avrich (1976, 30). This last promise was perhaps designed to intimidate the boyars rather than to actually rally the insurgents.

135. Quoted in Avrich (1976, 157).

136. Rodriguez (2009, 82).

137. Stalin (1924, 89).

138. Mao (1937, 85). Che Guevara (1961) parroted the need for "a press or mimeograph for newspapers" (84).

139. Taber (1965, 31–2).

140. Guevara (1961, 105).

141. Martowych (1950, 17).

142. Quoted in Serge (1951, 39).

143. Chehabi (2006, 189).

144. Lenin (1918, 67).

145. Nordstrom and Martin (1992, 202); but cf. Taber (1965, 104).

146. Eagleton (1963, 59).

147. Khachikyan (2010, 125).

148. Barr (2011, 217); Taber (1965, 111); Byrd and Miri (2018, 159); and Horne (1977, 133).

149. Biddle (2021, 149).

150. Von der Heydte (1972, 126).

151. Bolt (2012). Dabiq refers to an area in Syria related to the eschatology of Da'esh.

152. Hagopian (1974, 258).

153. Walt (1996, 25–8).

154. Ali Shariati (1979) described a successful ideology as "a united, coordinated, and firm visage" (15).

155. Shariati (1979, 39). Ostovar (2018) argued that ideology "underpins policies, shapes culture, and is the heart of identity. It is also central to expressions of political violence" (20). In contrast, Walt (1996, 8) argued that revolutionary ideology was not a principal element of the relationship between revolution and war, as his conception of ideology principally concerned state behavior and the perceptions other states may form of it.

156. Chehabi (2006).

157. Chehabi (2006, 231–2).

158. Shariati (1979, 41).

Chapter 9

1. Clausewitz (1832, 389).
2. Goodwin (2001, 40).
3. Zarrinkoub (1957, xxiii).
4. Kipling (1886). A *jezail* was a cheap rifle popular among South Asian insurgents resisting British imperialism.
5. Fukuyama (1992).
6. See Cialdini (1984) for more on social proof in terms of social influence.
7. E.g., North, Wallis, and Weingast (2013).
8. Fanon (1961, 22).
9. Cialdini and Goldstein (2004, 592).
10. In this context, Acemoglu and Robinson (2012a) described politics as "the process by which a society chooses the rules that will govern it" (79).
11. Quoted in Tourison (1991, 187).
12. Tilly (2003, 50).
13. Tilly (2003, 127).

14. Tilly (2003, 187).
15. Tilly (2003, 187).
16. Tilly (2003, 227).
17. Tilly (2003, 84).
18. Bayley and Perito (2011, 100).
19. Zonis (1983, 586–9).
20. See, e.g., Zimbardo (2007), Milgram (1963), Forsythe (2010), and Le Bon (1895).
21. Weber (1918).
22. Bernays (1928).
23. Gladwell (2000).
24. Martowych (1950).
25. Quoted in Alden, Thakur, and Arnold (2011, 65–86).
26. Acemoglu and Robinson (2012a, 3; 2019).
27. McNeill (1983, 320).
28. Gilio (1970, 13).
29. Shariati (1979, 39). Bronner (2017) observed that "Every mass ideology has a utopian component ... it is the ideal for which countless individuals have proven willing to die" (77).
30. Tainter (1988) noted that "Complex societies ... occupy every sector of the globe.... This is a new factor in human history. Complex societies as a whole are a recent and unusual aspect of human life. *The current situation, where all societies are so oddly constituted, is unique*" (213, emphasis in original).
31. Scott (2017, 117). But cf. Foa (2016), who disagreed. In any case, Knapp, et al. (2016) put the point well: "If Mesopotamia's long history lasted for an hour, then the nation-state has existed for only a second" (9).
32. Goodwin (2001, 30).
33. Alden, Thakur, and Arnold (2011, 154).
34. Amenta (2005, 104).
35. Both definitions are in Scott (2017, 9).
36. Carneiro (1977, 3).
37. Scott (2017, 118); Weber (1918).
38. Schelling (1960, 68).
39. Atabaki and Zürcher (2017, 1).
40. Scott (2017, 14).
41. Scott (2017, 125).
42. Scott (2017, 126, 235–53).
43. Arendt (1963, 192–3).
44. Mearsheimer (2001, 55). This implies that states with the largest militaries are drawn to use them, while states with the smallest militaries are drawn to avoid military adventures.
45. Fearon (2010) conceptualized sovereignty into a set of "practical arguments and conventions that emerge out of bargaining and bloody conflict between organizations that specialize in violence" (341).
46. Bayley and Perito (2010, 37).
47. Tainter (1988) and Cohen (1978, 4) both argued that there is actually no universal dividing attribute differentiating states from non-states, though I do not subscribe to that view.
48. Carré de Malberg (1920).
49. Quoted in Hassner (1989, 581–621).
50. Keohane (1984, xi).
51. Gilpin (1975); Keohane (1984, 18).
52. Acemoglu and Robinson (2012a, 238).
53. Perloff (2014, 19, 23) categorized the four types of social influence as persuasion, coercion, propaganda, and manipulation.
54. Carneiro (1977, 4).
55. Wittfogel (1957); (Carneiro 1977, 4).
56. Acemoglu and Robinson (2012a, 37).
57. Skocpol (1979, 95).
58. Bueno de Mesquita, et al. (2003, 474–5).
59. Goodwin (2001, 35).
60. Pittaway (2004, 6).
61. Carneiro (1977, 18).
62. Ramsey (2007). It is interesting to note here that Kipling's famous 1899 poem, "White Man's Burden," was in fact an anti-colonial poem about the U.S. counterinsurgency operations in the Philippines that emerged out of the Spanish-American War (Bew 2016, 107), despite the official U.S. Army history of U.S. military involvement there viewing it as simply a "campaign of benevolent pacification" (Birtle 2009, 120). See Ramsey (2007) for a discussion of this operation and for reproductions of the counterinsurgency orders U.S. Army Brigadier General Franklin Bell issued to his troops there.
63. Osanka (1962); Thayer (1963).
64. CIA (1977); Bender (1972).
65. Hersh (1991, 13).
66. Scott (1999, 23; 2017, 116).
67. Rawls (1971, 49).
68. Van den Berg and Janoski (2005, 84).
69. According to Waltz (1979), "force is least visible where power is most fully and adequately present. Power maintains

an order; the use of force signals a possible breakdown" (185).

70. Based on Locke (1690) and Waltz (1959, 58; 1979, 185).

71. Hovsepian (2008, 30–2).

72. Locke (1690, 416, 415).

73. For discussions of these behaviors, see Rosecrance and Stein (1993); Levy and Thompson (2010); and Waltz (1993).

74. Fearon (1995, 403).

75. See, e.g., Acemoglu and Robinson (2012a); Huntington (1957); Tilly (1990); McNeill (1983); Waltz (1959; 1979); and Durkheim (1912).

76. Jomini (1862, 27).

77. Clausewitz (1832, 23).

78. Fearon and Laitin (2003) concluded that "Civil war has been a far greater scourge than interstate war in this period, though it has been studied far less" (75).

79. Studies such as Chehab's (2005) analysis of the Iraq War after 2003 successfully employed an anatomy of causes approach in an irregular war context, providing fruitful results. Bueno de Mesquita, et al. (2003) observed that "The cause of or impetus for revolution is one thing. The consequences of such event are an entirely separate matter" (374).

80. Taber (1965, 151).

81. Mao (1937, 6).

82. Chehab (2005, 239).

83. Rodriguez (2009, 70, 81).

84. Rodriguez (2009).

85. Taber (1965, 170).

86. Acemoglu and Robinson (2012a, 372, 398). More broadly, North, Wallis, and Weingast (2013) noted that "Every explanation of large-scale social change contains a theory of economics, a theory of politics, and a theory of social behavior" (xvii).

87. Pollack (2019) alluded to this in his own conventional war study, reflecting that "To really understand the sources of military power, we often need to understand the societal factors that drive them" (xi).

88. Proponents who have argued for this "march of history" view include Hegel (1837); Marx (1844); Kojève (1957; 1993); and Fukuyama (1992).

89. See Lynn (1993) for more on the capacity to act among different actors in the conflict.

90. The tension between states and insurgencies competing for legitimacy capital is reflected in the state-centered description one analyst gave for resistance fighters in the Sahel-Maghreb, who he described as "loosely affiliated Libyan, Tunisian, and Egyptian jihadists" (Chivvis 2015, 5). That description begins from the false premise that the forces described held any feelings of loyalty toward the system that created the states of modern Libya, Tunisia, and Egypt, each a result of Italian, French, and British colonialism, respectively. For those jihadists, those states were neither legitimate nor rational.

91. The regime rationality axis was inspired by the discussion in Byrd and Miri (2018, 133).

92. Acemoglu and Robinson (2012a).

93. Acemoglu and Robinson (2012a, 66).

94. Scott (1998) observed that "Every revolution creates a temporary power vacuum when the power of the *ancien régime* has been destroyed but the revolutionary regime has not yet asserted itself throughout the territory" (206).

95. Tainter (1988, 4).

96. Alden, Thakur, and Arnold (2011, 154).

97. Tainter (1988, 213).

98. Acemoglu and Robinson (2012a, 106–7).

99. Traboulsi (2012, 70); Van den Berg and Janoski (2005, 80). Thayer (1963) similarly argued that "The cause must appear to be unachievable by less violent means. For only if the political or legal system seems to deny redress will the risks of violence be acceptable" (52).

100. Quoted in Horne (1977, 97).

101. Quoted in Horne (1977, 185).

102. Liddell Hart (1967, 358).

103. Rodriguez (2009, 23).

104. Musa and Horst (2019, 35).

105. Palmer (2018).

106. LeRiche and Arnold (2012, 124–5).

107. Trotsky (1930, 23).

108. Wagner (2007, 34).

109. Trotsky (1930, 1).

110. Chehabi (2006, 228).

111. In their analysis of unrestricted warfare in a potential conflict between the U.S. and China, Qiao and Wang (1999) observed that "The most modern military force does not have the ability to control public clamor and cannot deal with an opponent who does things in an unconventional manner" (13).

112. Waltzer (1977); Bauer (2008); and Couzigou (2017).
113. Gordon and Trainor (1995).
114. Clark (2002); Lamb (2018); and Fogarty Report (1988).
115. Pollack (2019, 28–9, 487); Hackett (2018, 20).
116. Mao (1937, 93).
117. Qiao and Wang (1999, 16).
118. Engels (1880, 16).
119. Machiavelli (1531, 6). The importance of having popular support was hardly a novel concept even during Machiavelli's era. Chinese military theorist Li Ch'uan (fl. 8th century CE) warned that "When an army penetrates far into enemy territory, care must be taken not to alienate the people," so that "the hearts of all may be won" (quoted in Sun Tzu, 167).
120. Machiavelli (1531, 8, 69).
121. Quoted in (Alleg 1958, xlii).
122. The concept of selective violence also operates within this model. See Kalyvas (2006, 196) for more on selective violence.
123. Rogers (2016) made the important observation that "Too often, thinking about civil war and particularly about the connections between civil war and the state has little to no analysis of what civil war is and what may characterize phenomena labelled in this way" (7–8).
124. Alden, Thakur, and Arnold (2011, xi).
125. Chakrabarti (2010, 78).
126. Nagl (2005, 128).
127. Alden, Thakur, and Arnold (2011, 15).
128. Trinquier (1964, 74).
129. Trinquier (1964, 53).

Concluding Irregular War

1. Jomini (1862, 165).
2. See Schelling (1960, 74).
3. Schelling (1960, 91).
4. Mearsheimer (2001, 409).
5. Quoted in Atabaki and Zürcher (2017, 183).
6. Walt (1996) is an example of the blind spot in scholarship on irregular war. He asserted "War is ultimately a response to problems that arise between two or more states" (12), ignoring completely the more prevalent wars within them.
7. Mearsheimer (2001, 408). Waltz (1979) observed that "The most destructive wars of the hundred years following the defeat of Napoleon took place not among states but *within* them…. We easily lose sight of the fact that struggles to achieve and maintain power, to establish order, and to contrive a kind of justice within states may be bloodier than wars among them" (103, emphasis in original).
8. Kalyvas (2001); Kalyvas and Kocher (2009).
9. Kalyvas and Kocher (2009, 336).
10. Geddes (2009, 4).
11. E.g., Fearon (2010) and Wagner (2007).
12. Fearon (2010) optimistically remarked that, "the field remains wide open for foundational work on central questions" (336) due to the lack of sound theory.
13. Tainter (1988, 54).
14. See Clausewitz (1832, 546) for his brief mention of the ends of war.

References

Acemoglu, Daron, and James Robinson. 2012a. *Economic Origins of Dictatorship and Democracy*. 8th ed. New York: Cambridge University Press.

Acemoglu, Daron, and James Robinson. 2012b. *Why Nations Fail: The Origins of Power, Prosperity, and Poverty*. New York: Crown Business.

Acemoglu, Daron, and James Robinson. 2019. *The Narrow Corridor: States, Societies, and the Fate of Liberty*. New York: Penguin.

Adamsky, Dimitry. 2015. *Cross-Domain Coercion: The Current Russian Art of Strategy*. Paris: IFRI Security Studies Center.

Alden, Chris, Monika Thakur, and Matthew Arnold. 2011. *Militias and the Challenges of Post-Conflict Peace: Silencing the Guns*. London: Zed Books.

Aliyev, Huseyn. 2016. "Strong Militaries, Weak States, and Armed Violence." *Security Dialogue* 47, no. 6: 498–516.

Alleg, Henri. 1958. *Le Question*. Translated by John Calder. Paris: Minuit.

Allison, Christine. 2001. *The Yezidi Oral Tradition in Iraqi Kurdistan*. New York: Routledge.

Almond, Gabriel. 1991. "Capitalism and Democracy." *American Political Science Association* 24, no. 3: 467–74.

Alon, Yoav. 2006. "The Balqa Revolt: Tribes and Early State-Building in Transjordan." *Die Welt des Islams* 46, no. 1: 7–42.

Amanat, Abbas. 2017. *Iran: A Modern History*. New Haven: Yale University Press.

Amenta, Edwin. 2005. "State-Centered and Political Institutional Theory: Retrospect and Prospect." In *The Handbook of Political Sociology: States, Civil Societies, and Globalization*, edited by Thomas Janoski, Robert Alford, Alexander Hicks, and Mildred Schwartz, 96–114. New York: Cambridge University Press.

Arendt, Hannah. 1951. *The Origins of Totalitarianism*. New York: Harcourt.

Arendt, Hannah. [1963] 2006. *On Revolution*. London: Penguin.

Arquilla, John. 2011. *Insurgents, Raiders, and Bandits*. Lanham, MD: Ivan R. Dee.

Arquilla, John. 2018. "Perils of the Gray Zone: Paradigms Lost, Paradoxes Regained." *Prism* 7, no. 3: 118–29.

Atabaki, Touraj, and Erik Zurcher, eds. 2017. *Men of Order: Authoritarian Modernization Under Ataturk and Reza Shah*. London: I.B. Tauris.

Avrich, Paul. 1976. *Russian Rebels, 1600–1800*. New York: Norton Library.

Azizi, Arash. 2020. *The Shadow Commander: Soleimani, the US, and Iran's Global Ambitions*. London: One World.

Azzam, Abdullah. 1988. *Join the Caravan*. N.p.

Baker, James, Lee Hamilton, Lawrence Eagleburger, Vernon Jordan, Edwin Meese, Sandra Day O'Connor, Leon Panetta, William Perry, Charles Robb, and Alan Simpson. 2006. *The Iraq Study Group Report: The Way Forward—A New Approach*. New York: Vintage.

Baniyaghoob, Jila. 2013. *Women of Evin: Ward 209*. Middletown, DE: Xlibris.

Barber, W.H. 1951. "Patriotism and 'Gloire' in Corneille's 'Horace.'" *The Modern Language Review* 46, no. 3/4: 368–78.

Barbera, Salvador, and Matthew Jackson. 2020. "A Model of Protests, Revolution, and Information." *Quarterly Journal of Political Science* 15, no. 3: 297–335.

Barnes, Harry Elmer. 1924. "Theories of

the Origin of the State in Classical Political Philosophy." *The Monist* 34, no. 1: 15–62.

Barr, James. 2011. *A Line in the Sand*. London: Simon & Schuster.

Bassiouni, M. Cherif. 2008. "The New Wars and the Crisis of Compliance with the Law of Armed Conflict by Non-State Actors." *The Journal of Criminal Law and Criminology* 98, no. 3: 711–810.

Bastiat, Frederic. [1850] 2007. *The Law*. Auburn, AL: Tribeca Books.

Bates, Robert, Avner Greif, Margaret Levi, Jean-Laurent Rosenthal, and Barry Weingast. 2000. "The Analytical Narrative Project." *American Political Science Review* 94, no. 3: 696–702.

Bauer, John. 2008. "Justice: A Problem for Military Ethics during Irregular War." Master Thesis. Fort Leavenworth: U.S. Army Command and General Staff College.

Bayley, David, and Robert Perito. 2010. *The Police in War: Fighting Insurgency, Terrorism, and Violent Crime*. London: Lynne Rienner.

Bechara, Souha. 2003. *Resistance: My Life for Lebanon*. New York: Soft Skull Press.

Bell, Kevin. 2016. "The First Islamic State: A Look Back at the Islamic State of Kunar." *CTC Sentinel* 9, no. 2: 10–13.

Bender, Gerald. 1972. "The Limits of Counterinsurgency: An African Case." *Comparative Politics* 4, no. 3: 331–60.

Bennett, Huw. 2013. *Fighting the Mau Mau: The British Army and Counter-Insurgency in the Kenya Emergency*. Cambridge: Cambridge University Press.

Berger, D.H. 2020. *Marine Corps Doctrinal Publication 7—Learning*. Washington, D.C.: Headquarters, U.S. Marine Corps.

Bernays, Edward. [1928] 2005. *Propaganda*. Brooklyn: Ig.

Bew, John. 2016. *Realpolitik: A History*. Oxford: Oxford University Press.

Bezins, Janis. 2014. "Russia's New Generation Warfare in Ukraine: Implications for Defense Policy." *Military Operations* 2, no. 4: 4–7.

Bickel, Keith. 2004. "(Review) Strategy for Chaos: Revolutions in Military Affairs and the Evidence of History." *The Journal of Military History* 68, no. 1: 321–22.

Bickerton, Ian, and Carla Klausner. 2015. *A History of the Arab-Israeli Conflict*. 7th ed. Boston: Pearson.

Biddle, Stephen. 2021. *Nonstate Warfare: The Military Methods of Guerillas, Warlords, and Militias*. Princeton: Princeton University Press.

Bill, James. 1988. *The Eagle and the Lion: The Tragedy of American-Iranian Relations*. New Haven: Yale University Press.

Birtle, Andrew. 2007. *U.S. Army Counterinsurgency and Contingency Operations Doctrine, 1942–1976*. Washington, D.C.: Center of Military History, U.S. Army.

Birtle, Andrew. 2009. *U.S. Army Counterinsurgency and Contingency Operations Doctrine, 1860–1941*. Washington, D.C.: Center of Military History, U.S. Army.

Boatwright, Mary, Daniel Gargola, and Richard Talbert. 2004. *The Romans: From Village to Empire*. New York: Oxford University Press.

Bolt, Neville. 2012. *The Violent Image: Insurgent Propaganda and the New Revolutionaries*. New York: Columbia University Press.

Bonaparte, Napoleon. [1827] 1902. *Military Maxims*. Translated by G.C. D'Aguilar. Paris: n.p.

Bou-Nacklie, N.E. 1993. "Les Troupes Speciales: Religious and Ethnic Recruitment, 1916–46." *International Journal of Middle East Studies* 25, no. 4: 645–60.

Bourcet, Pierre-Joseph. 1775. *Principes de la Guerre de Montagnes [Principles of Mountain Warfare]*. Paris: Imprimerie Nationale.

Brafman, Ori, and Rod Beckstrom. 2006. *The Starfish and the Spider*. New York: Penguin.

Broadwell, Paula. 2012. *All In: The Education of General David Petraeus*. New York: Penguin.

Bronner, Stephen. 2017. *Critical Theory: A Brief Introduction*. Oxford: Oxford University Press.

Brubaker, Rogers. 2006. *Ethnicity without Groups*. Cambridge: Harvard University Press.

Bueno de Mesquita, Bruce, Alastair Smith, Randolph Siverson, and James Morrow. 2003. *The Logic of Political Survival*. Cambridge: MIT Press.

Bueno de Mesquita, Bruce, Randolph Siverson, and Gary Woller. 1992. "War and the Fate of Regimes: A Comparative Analysis." *The American Political Science Review* 86, no. 3: 638–46.

Burke, Edmund. [1790] 2008. *The Evils of Revolution*. London: Penguin.

Byely, B., Y. Dzyuba, G. Fyodorov, Y. Khomenko, T. Kondratkov, V. Kozlov Kulakov, Y. Medvedev, et al. 1972. *Marksizm-Leninizm o voine i armii [Marxism-Leninism on war and army]*. Moscow: Progress.

Byman, Daniel. 2007. *Understanding Proto-Insurgencies*. Santa Monica: RAND.

Byrd, Dustin, and Javad Miri, eds. 2018. *Ali Shariati and the Future of Social Theory: Religion, Revolution, and the Role of the Intellectual*. Chicago: Haymarket Books.

Caballero, Ricardo, and Mohamad Hammour. 2000. "Creative Destruction and Development: Institutions, Crises, and Restructuring." *NBER Working Paper*, no. 7849.

Callwell, Charles. 1896. *Small Wars: Their Principles and Practice*. London: Harrison and Sons.

Carneiro, Robert. 1977. "A Theory of the Origin of the State." *Studies in Social Theory*, no. 3: 3–21.

Carré de Malberg, Raymond. 1920. *Contribution à la théorie générale de l'Etat [Contribution to the general theory of the state]*. Paris: Librairie de la Société du Recueil Sirey.

Carson, Austin. 2018. *Secret Wars: Covert Conflict in International Politics*. Princeton: Princeton University Press.

Castelli, Christopher. 2009. "'Irregular Warfare' Term Stirs Debate as DOD. Prepares for QDR." *Inside the Pentagon* 25, no. 15: 18–19.

Caverley, Jonathan. 2010. "Explaining U.S. Military Strategy in Vietnam: Thinking Clearly about Causation." *International Security* 35, no. 3: 124–43.

Celeski, Joseph. 2010. "Hunter Killer Teams: Attacking Enemy Safe Havens." *JSOU Report* 10-1: 1–82.

Central Intelligence Agency. 1977. *Soviet and Cuban Intervention in the Angolan Civil War*. Washington, D.C.: Central Intelligence Agency.

Central Treaty Organization. 1955. *The 'Baghdad Pact.'* Baghdad: CETO.

Chakrabarti, Shantanu. 2010. "Evolving Insurgency and India's Counterinsurgency Options: Entering into the Age of Fourth-Generation Warfare?" *The Quarterly Journal*: 65–78.

Chandler, David. 1989. "Introduction: Regular and Irregular Warfare." *The International History Review* 11, no. 1: 2–13.

Chehab, Zaki. 2005. *Inside the Resistance: The Iraqi Insurgency and the Future of the Middle East*. New York: Nation Books.

Chehab, Zaki. 2007. *Inside Hamas: The Untold Story of the Militant Islamic Movement*. New York: Nation Books.

Chehabi, H.E, ed. 2006. *Distant Relations: Iran and Lebanon in the Last 500 Years*. London: I.B. Tauris.

Chekinov, S.G., and S. Bogdanov. 2013. *The Nature and Content of a New-Generation War*. Moscow: n.p.

Chivvis, Christopher. 2014. *Toppling Qaddafi: Libya and the Limits of Liberal Intervention*. New York: Cambridge University Press.

Chivvis, Christopher. 2015. *The French War on al-Qaeda*. New York: Cambridge University Press.

Cialdini, Robert. 1984. *Influence: The Psychology of Persuasion*. New York: HarperCollins.

Cialdini, Robert, and Noah Goldstein. 2004. "Social Influence: Compliance and Conformity." *Annual Review of Psychology* 55: 591–621.

Clark, Wesley. 2002. *Waging Modern War*. Cambridge: Perseus.

Cleveland, Charles, and Daniel Egel. 2020. *The American Way of Irregular War*. Santa Monica: RAND.

Clutterbuck, Richard. 1980. *Guerrillas and Terrorists*. Columbus: Ohio University Press.

Cohen, Ronald. 1978. *Origins of the State: The Anthropology of Political Evolution*. Philadelphia: Institute for the Study of Human Issues.

Cohen, Ronald. 1984. "Warfare and State Formation: Wars Make States and States Make Wars." In R. Brian Ferguson, ed., *Warfare, Culture, and Environment*, 464–70. Orlando: Academic Press.

Copeland, Thomas, ed. 1958. *The Correspondence of Edmund Burke*. Chicago: University of Chicago Press.

Cordesman, Anthony. 2017. "Stability Operations in Syria: The Need for a Revolution in Civil-Military Affairs." *Military Review Online Exclusive*: 2–21.

Corum, James. 2006. *Training Indigenous*

Forces in Counterinsurgency: A Tale of Two Insurgencies. Carlisle, PA: Strategic Studies Institute.

Couzigou, Irene. 2017. "The Fight against the 'Islamic State' in Syria." *Geopolitics, History, and International Relations* 9, no. 2: 80–106.

Crawford, Neta. 2019. *United States Budgetary Costs and Obligations of Post-9/11 Wars through FY2020: $6.4 Trillion.* Providence: Brown University Costs of War Project.

Creveld, Martin van. [1977] 2014. *Supplying War: Logistics from Wallenstein to Patton.* New York: Cambridge University Press.

Creveld, Martin van. 1986. *Technology and War: From 2000 B.C. to the Present.* New York: Free Press.

Creveld, Martin van. 1991. *The Transformation of War.* New York: Simon & Schuster.

Crossett, Chuck, ed. 2012. *Casebook on Insurgency and Revolutionary Warfare Volume II: 1962-2009.* Fort Bragg, NC: USASOC.

Dawkins, Richard. [1976] 2006. *The Selfish Gene.* Oxford: Oxford University Press.

Department of Defense. 2020. "Summary of the Irregular Warfare Annex to the National Defense Strategy." Office of the Secretary of Defense. Washington, D.C.: Government Printing Office.

Department of State. 2006. "1207 Funding." Office of the Coordinator for Reconstruction and Stabilization. Retrieved from https://2001-2009.state.gov/s/crs/107030.htm.

Department of the Army. 1982. *Field Manual 34-71: North Korean Military Forces.* Headquarters, Department of the Army.

Department of the Army. 1984. *Field Manual 100-2-2: Soviet Specialized Warfare and Rear Area Support.* Headquarters, Department of the Army.

Department of the Army. 1991. *NTC Handbook 100-91: The Iraqi Army—Organization and Tactics.* National Training Center.

Department of the Army. 2006. *Counterinsurgency Field Manual 3-24/MCWIP 3-33.5.* Washington, D.C.: Government Printing Office.

Department of the Army. 2007. *(S//NF) Special Forces Unconventional Warfare Operations Field Manual 3-05.201 (U).* Washington, D.C.: Government Printing Office.

Department of the Army. 2008. *Army Special Forces Unconventional Warfare Operations Field Manual 3-05.130.* Washington, D.C.: Government Printing Office.

Department of the Army. 2015. *ATP 3-18.20 (S//NF) Advanced Special Operations Techniques (ASOT) (U).* Washington, D.C.: Government Printing Office.

Department of the Navy. 1940. *Small Wars Manual.* Washington, D.C.: Government Printing Office.

De Soto, Hernando. 2001. "The Mystery of Capital: Finance and Development." *Finance and Development* 39, no. 1: 29–33.

Devlin, Larry. 2007. *Chief of Station: Congo.* Philadelphia: Perseus.

De Waal, Thomas. 2019. *The Caucasus: An Introduction.* Oxford: Oxford University Press.

Diamond, Martin. 1987. "The Federalist 1787-1788." In *History of Political Philosophy*, 3rd ed., edited by Leo Strauss and Joseph Cropsey, 659–79. Chicago: University of Chicago Press.

Didier, Leroy, and Elena Aoun. 2019. "Crossed Views on Jihadism in the Middle East: The Engagement of Lebanese Fighters in Syria." *Security and Strategy*, no. 140.

Dingel, Eva. 2013. "Hezbollah's Rise and Decline? How the Political Structure Seems to Harness the Power of Lebanon's Non-State Armed Group." *Security and Peace* 31, no. 2: 70–76.

Dittgen, Herbert. 1999. "World without Borders? Reflections on the Future of the Nation-State." *Government and Opposition* 34, no. 2: 161–79.

Dixit, K.C. 2010. "Sub-Conventional Warfare Requirements, Impact and Way Ahead." *Journal of Defence Studies* 4, no. 1: 120–34.

Doerr, Arthur, Jerome Coling, and William Kerr. 1970. "Agricultural Evolution in Israel in the Two Decades since Independence." *Middle East Journal* 24, no. 3: 319–37.

Durkheim, Emile. 1893. *De la division du travail social: étude sur l'organisation des sociétés supérieures [Of the Division of Labor in Society: A Study on the Organization of Higher Societies].* Paris: Alcan.

Durkheim, Emile. [1912] 2008. *The Elementary Forms of Religious Life.* Translated by Carol Cosman. Oxford: Oxford University Press.

Eagleton, William. 1963. *The Kurdish Republic of 1946.* Oxford: Oxford University Press.

Echevarria, Antulio. 2005. *Fourth-Generation War and Other Myths.* Carlisle, PA: SSI Publications.

Eisenstadt, Michael, and Kenneth Pollack. 2020. "Training Better Arab Armies." *Parameters* 50, no. 3: 96–111.

El-Sisi, Abdel Fattah. 2006. "Democracy in the Middle East." Master Thesis. Carlisle Barracks, PA: U.S. Army War College.

Emmerich, A. 1789. *The Partisan in War, of the use of a Corps of Light Troops to an Army.* London: Reynell.

Engels, Frederick. 1880. *Socialism, Utopian or Scientific.* Translated by Edward Aveling. Paris: Revue Socialiste.

Erasmus, Desiderius. [1512] 1979. *Encomium Moriae [The Praise of Folly].* New Haven: Yale University Press.

Evans, Michael. 2005. "Elegant Irrelevance Revisited: A Critique of Fourth-Generation Warfare." *Contemporary Security Policy* 26, no. 2: 242–49.

Fagan, Brian. 2000. *The Little Ice Age: How Climate Made History 1300–1850.* New York: Basic Books.

Fanon, Frantz. [1961] 2004. *Les damnes de la terre [The Wretched of the Earth].* Translated by Richard Philcox. New York: Grove Press.

Fearon, James. 1995. "Rationalist Explanations for War." *International Organization* 49, no. 3: 379–414.

Fearon, James. 2010. "Comments on R. Harrison Wagner's War and the State: The Theory of International Politics." *International Theory* 2, no. 2: 333–43.

Fearon, James, and David Laitin. 2003. "Ethnicity, Insurgency, and Civil War." *The American Political Science Review* 97, no. 1: 75–90.

Fernández, Walter, Michael Martin, Shirley Gregor, Steven Stern, and Michael Vitale. 2007. "A Multi-Paradigm Approach to Grounded Theory." In *Information Systems Foundations: Theory, Representation and Reality,* edited by Dennis Hart and Shirley Gregor, 231–46. Canberra: ANU Press.

Fiala, Otto. 2020. *Resistance Operating Concept (ROC).* MacDill Air Force Base, FL: The JSOU Press.

Fields, Gary. 2017. *Enclosure: Palestinian Landscapes in a Historical Mirror.* Oakland: University of California Press.

Finkel, Caroline. 2005. *Osman's Dream: The History of the Ottoman Empire.* New York: Basic Books.

Fishel, John, and Max Manwaring. 2008. *Uncomfortable Wars Revisited.* Oklahoma City: University of Oklahoma Press.

Foa, Roberto. 2016. "Ancient Polities, Modern States." Doctoral Dissertation. Cambridge: Harvard University, Graduate School of Arts and Sciences.

Fogarty, William. 1988. *Formal Investigation into the Circumstances Surrounding the Downing of Iran Air Flight 655 on 3 July 1988.* Washington, D.C.: Department of Defense.

Forbes, Hugh. 1855. *Manual for the Patriotic Volunteer on Active Service in Regular and Irregular War: Being the Art and Science of Obtaining and Maintaining Liberty and Independence, Volume 1.* New York: W.H. Tinson.

Forgas, J.P., and K.D. Williams, eds. 2001. *Social Influence: Direct and Indirect Processes.* Philadelphia: Routledge.

Forsythe, Donelson. 2010. *Group Dynamics.* Boston: Cengage.

Forth, Aidan, and Jonas Kreienbaum. 2016. "A Shared Malady." *Journal of Modern European History* 14, no. 2: 245–67.

Fravel, M. Taylor. 2019. *Active Defense: China's Military Strategy Since 1949.* Princeton: Princeton University Press.

French Jr., J.R.P., and B.H. Raven. 1959. "The bases of social power." In *Studies in Social Power,* edited by D. Cartwright, 150–67. Ann Arbor: Institute for Social Research.

Fuccaro, Nelida. 1999. "Communalism and the State in Iraq: The Yazidi Kurds, c. 1869–1940." *Middle Eastern Studies* 35, no. 2: 1–26.

Fukuyama, Francis. 1992. *The End of History and the Last Man.* New York: Simon & Schuster.

Fukuyama, Francis. 2012. *The Origins of Political Order.* New York: Farrar, Straus, and Giroux.

Fukuyama, Francis. 2015. *Political Order and Political Decay.* New York: Farrar, Straus, and Giroux.

Fukuyama, Francis, ed. 2007. *Blindside: How to Anticipate Forcing Events and Wild Cards in Global Politics*. Washington, D.C.: Brookings Institution Press.

Gaddis, John Lewis. 2002. "On Strategic Surprise." *Hoover Digest*, no. 2.

Galeotti, Mark. 2019. *Armies of Russia's War in Ukraine*. Oxford: Osprey.

Galula, David. [1964] 2006. *Counterinsurgency Warfare: Theory and Practice*. New York: Praeger.

Gandhi, Mahatma. 1961. *Non-Violent Resistance (Satyagraha)*. New York: Schocken Books.

Garrett-Peltier, Heidi. 2019. *War Spending and Lost Opportunities*. Providence: Brown University Costs of War Project.

Geddes, Barbara. 2009. *Paradigms and Sand Castles: Theory Building and Research Design in Comparative Politics*. Ann Arbor: University of Michigan Press.

Geertz, Clifford. 1973. *The Interpretation of Cultures*. New York: Basic Books.

Gellner, Ernst. 1988. *Plough, Sword, and Book*. Chicago: Chicago University Press.

Ghaddar, Hanin, ed. 2018. *Iran's Foreign Legion: The Impact of Shia Militias on U.S. Foreign Policy, Policy Note 46*. Washington, D.C.: Washington Institute for Near East Policy.

Ghazvinian, John. 2021. *America and Iran, a History: 1720 to the Present*. New York: Knopf Doubleday.

Giap, Vo Nguyen. 1970. *The Military Art of the People's War: Selected Writings of General Vo Nguyen Giap*. Edited by Russell Stetler. New York: Monthly Review Press.

Gilio, Maria Esther. [1970] 1973. *La guerrilla tupamara [The Tupamaro Guerrillas]*. Translated by Anne Edmonson. New York: Ballantine.

Gilpin, Robert. 1975. *U.S. Power and the Multinational Corporation*. New York: Basic Books.

Gladwell, Malcom. 2000. *The Tipping Point: How Little Things Can Make a Big Difference*. New York: Back Bay Books.

Glahn, Gerhardt von. 1957. *The Occupation of Enemy Territory: A commentary on the law and practice of belligerent occupation*. Minneapolis: University of Minnesota Press.

Glaser, Barney, and Anselm Strauss. 1967. *The Discovery of Grounded Theory: Strategies for Qualitative Research*. Piscataway: Rutgers—The State University.

Glavin, Terry. 2015. "No Friends but the Mountains: The Fate of the Kurds." *World Affairs* 177, no. 6: 57–66.

Goldenziel, Jill. 2020. "Law as a Battlefield: The U.S., China, and Global Escalation of Lawfare." *Cornell Law Review* 106: 1085–172.

Goldstone, Jack. 2014. *Revolutions: A Very Short Introduction*. New York: Oxford University Press.

Goodwin, Jeffrey. 2001. *No Other Way Out: States and Revolutionary Movements, 1945–1991*. New York: Cambridge University Press.

Goodwin, Jeffrey. 2005. "Revolutions and Revolutionary Movements." In *The Handbook of Political Sociology: States, Civil Societies, and Globalization*, edited by Thomas Janoski, Robert Alford, Alexander Hicks, and Mildred Schwartz, 404–22. New York: Cambridge University Press.

Goor, Jurrien van. 2004. *Prelude to Colonialism: The Dutch in Asia*. Netherlands: Uitgeverij Verloren.

Gordon, Michael, and Bernard Trainor. 1995. *The General's War*. New York: Back ay Books.

Govia, J.M. 1994. "The Psychological Profile of Survival, Evasion, Resistance, and Escape Instructor Personnel." Technical Report. Bethesda, MD: National Naval Medical Center.

Gray, Alfred. 1989. *Fleet Marine Force Manual—Warfighting*. Washington, D.C.: Headquarters, U.S. Marine Corps.

Gray, Colin. 2002. *Strategy for Chaos: Revolutions in Military Affairs and the Evidence of History*. Portland, OR: Frank Cass.

Gray, Colin. 2005. *Transformation and Strategic Surprise*. Carlisle, PA: SSI Publications.

Gray, Colin. 2006. *Irregular Enemies and the Essence of Strategy: Can the American Way of War Adapt?* Carlisle, PA: SSI Publications.

Guevara, Ernesto "Che." [1961] 1998. *Guerrilla Warfare*. Edited by Marc Becker. Lincoln: University of Nebraska Press.

H. 1956. *The Handbook for Irish Volunteers: Simple Lectures on Military Subjects*. Dublin: n.p.

Hackett, Jonathan. 2018. "Magna

Phoenicia: Elevating Phoenician Civilization in Ancient History." Unpublished Master Thesis. Charles Town, WV: American Public University System.

Hackett, Jonathan. 2021. "Toward a Theory of Irregular War." Unpublished Master Thesis. Charles Town, WV: American Public University System.

Hagopian, Mark. 1974. *The Phenomenon of Revolution*. New York: Dodd, Mead.

Hammes, Thomas. 2006. *The Sling and the Stone: On War in the 21st Century*. St. Paul: Zenith Press.

Hammond, Andrew. 2015. "Through a Glass, Darkly: The CIA and Oral History." *History* 100, no. 2: 311–26.

Hari, Daoud. 2008. *The Translator*. New York: Random House.

Hassner, Pierre. 1987. "Immanuel Kant 1724–1804." In *History of Political Philosophy*, 3rd ed., edited by Leo Strauss and Joseph Cropsey, 581–621. Chicago: University of Chicago Press.

Hawkins, John. 2013. "The Limits of Fire Support: American Finances and Firepower Restraint during the Vietnam War." PhD Diss., Texas A&M University, College Station.

Heacock, Ashley. 2013. *Understanding Mali: Connections and Confrontations between the Tuareg, Islamist Rebels, and the Government*. N.p.

Hegel, Georg F.W. 1807. *Phaenomenologie des Geistes [The Phenomenology of Spirit]*. Bamburg und Wurzburg: Joseph Anton Goebhardt.

Hegel, Georg F.W. 1820. *Outlines of the Philosophy of Right*. Translated by T.M. Knox. Edited by Stephen Houlgate. Oxford: Oxford University Press.

Hegel, Georg F.W. 1837. *The Philosophy of History*. Edited by Eduard Gans. Berlin: n.p.

Helman, Christopher, and Hank Tucker. 2021. "The War in Afghanistan Cost America $300 Million Per Day for 20 Years, with Big Bills Yet To Come." *Forbes*. Retrieved from https://www.forbes.com/sites/hanktucker/2021/08/16/the-war-in-afghanistan-cost-america-300-million-per-day-for-20-years-with-big-bills-yet-to-come/.

Helvey, Robert. 2004. *On Strategic Nonviolent Conflict: Thinking about the Fundamentals*. Boston: Albert Einstein Foundation.

Hemingway, Ernest. 1940. *For Whom the Bell Tolls*. New York: Charles Scribner's Sons.

Henriksen, Thomas. 2007. "The Israeli Approach to Irregular Warfare and Implications for the United States." *JSOU Report* 07-3: 1–58.

Henriksen, Thomas. 2010. "Afghanistan, Counterinsurgency, and the Indirect Approach." *JSOU Report* 10-30: 1–92.

Hersh, Seymour. 1991. *The Samson Option*. New York: Random House.

Herz, John. 1950. "Idealist Internationalism and the Security Dilemma." *World Politics* 2, no. 2: 157–80.

Hildinger, Erik. 2002. *Swords against the Senate: The Rise of the Roman Army and the Fall of the Republic*. Cambridge: Perseus.

Hitler, Adolf. 1925. *Mein Kampf; Die Nationalsozialistiche Bewegung*. Munich: Franz Eher Nachfolger GmbH.

Hobbes, Thomas. [1651] 1985. *Leviathan or the Matter, Forme and Power of a Common-Wealth Ecclesiastical and Civill*. Edited by C.B. Macpherson. New York: Penguin.

Holloway, James, Samuel Wilson, Leroy Manor, James Smith, John Piotrowsky, and Alfred Gray. 1980. *Holloway Commission Report*. Department of Defense.

Hooks, Gregory, and James Rice. 2005. "War, Militarism, and States." In *The Handbook of Political Sociology: States, Civil Societies, and Globalization*, edited by Thomas Janoski, Robert Alford, Alexander Hicks, and Mildred Schwartz, 568–71. New York: Cambridge University Press.

Horne, Alistair. [1977] 2006. *A Savage War of Peace: Algeria 1954–1962*. New York: New York Review of Books.

Hovsepian, Nubar, ed. 2008. *The War on Lebanon: A Reader*. Northampton, MA: Olive Branch Press.

Howell, Kerry. 2015. "Grounded Theory." In *An Introduction to the Philosophy of Methodology*, 131–53. London: SAGE.

Hunter, Gordon M. 2005. "Qualitative Research in Information Systems: Consideration of Selected Theories." In *Information Systems Foundations: Construction and Criticising*, edited by Dennis Hart and Shirley Gregor, 35–40. Canberra: ANU Press.

Huntington, Samuel. [1957] 2003. *The*

Soldier and the State. Cambridge: Harvard University Press.
Huntington, Samuel. [1968] 2006. *Political Order and Changing Societies*. New Haven: Yale University Press.
Huntington, Samuel. 1993. "A Clash of Civilizations?" *Foreign Affairs* 72, no. 3: 22–49.
Ibn Khaldun. [1377] 2005. *Al-Muqaddimah [Prolegomena]*. Translated by Franz Rosenthal. Edited by N.J. Dawood. Princeton: Princeton University Press.
Inkles, Alex, and David Smith. 1974. *Becoming Modern: Individual Change in Six Developing Countries*. Cambridge: Harvard University Press.
Irwin, Will. 2019. "Support to Resistance: Strategic Purpose and Effectiveness." *JSOU Report* 19-2: 1–250.
Ismail, Salwa. 2018. *The Rule of Violence: Subjectivity, Memory and Government in Syria*. Cambridge: Cambridge University Press.
Jabbra, Joseph, and Nancy Jabbra. 1983. "Lebanon: Gateway to Peace in the Middle East?" *International Journal* 38, no. 4: 577–612.
Jalali, Ali, and Lester Grau. 1999. *The Other Side of the Mountain: Mujahideen Tactics in the Soviet-Afghan War*. Quantico, VA: USMC Studies and Analysis Division.
James, Paul. 2006. *Globalism, Nationalism, Tribalism: Bringing Theory Back In*. London: Sage.
Jasper, James. 2005. "Culture, Knowledge, and Politics." In *The Handbook of Political Sociology: States, Civil Societies, and Globalization*, edited by Thomas Janoski, Robert Alford, Alexander Hicks, and Mildred Schwartz, 115–34. New York: Cambridge University Press.
Jefferson, Thomas. 1776. *The Declaration of Independence*. Philadelphia: n.p.
Jentzsch, Corinna, Stathis Kalyvas, and Livia Schubiger. 2015. "Militias in Civil Wars." *The Journal of Conflict Resolution* 59, no. 5: 755–69.
Jervis, Robert. 1978. "Cooperation under the Security Dilemma." *World Politics* 30, no. 2: 167–214.
Joint Chiefs of Staff. 2006. *Irregular Warfare Joint Operating Concept 1.0*. Washington, D.C.: Government Printing Office.
Joint Chiefs of Staff. 2007. *Irregular Warfare Joint Operating Concept Revised Appendix C*. Washington, D.C.: Government Printing Office.
Joint Chiefs of Staff. 2010. *Irregular Warfare: Countering Irregular Threats—Joint Operating Concept 2.0*. Washington, D.C.: Government Printing Office.
Joint Chiefs of Staff. 2014a. *Joint Publication 1-02: Department of Defense Dictionary of Military and Associated Terms*. Washington, D.C.: Government Printing Office.
Joint Chiefs of Staff. 2014b. *Joint Publication 1-0: Doctrine of the United States Military*. Washington, D.C.: Government Printing Office.
Joint Chiefs of Staff. 2014c. *Joint Publication 3-05: Special Operations*. Washington, D.C.: Government Printing Office.
Joint Chiefs of Staff. 2017. *Joint Publication 3-20: Security Cooperation*. Washington, D.C.: Government Printing Office.
Joint Chiefs of Staff. 2019a. *Joint Doctrine Note 1-19: Competition Continuum*. Washington, D.C.: Government Printing Office.
Joint Chiefs of Staff. 2019b. *Joint Doctrine Note 2-19: Strategy*. Washington, D.C.: Government Printing Office.
Jomini, Antoine-Henri. [1862] 2008. *The Art of War*. Translated by G.H. Mendell and W.P Craighill. Radford, VA: Wilder Publications.
Jones, Jeremy, and Nicholas Rideout. 2015. *A History of Modern Oman*. New York: Cambridge University Press.
Jones, Seth. 2017. *Waging Insurgent Warfare*. Oxford: Oxford University Press.
Jonsson, Oscar. 2019. *The Russian Understanding of War: Blurring the Lines between War and Peace*. Washington, D.C.: Georgetown University Press.
Jung, Carl. [1957] 2002. *The Undiscovered Self*. Translated by Kegan Paul. London: Routledge Classics.
Kalyvas, Stathis. 1999. "Wanton and Senseless? The Logic of Massacres in Algeria." *Rationality and Society* 11, no. 3: 243–85.
Kalyvas, Stathis. 2001. "'New' and 'Old' Civil Wars: A Valid Distinction?" *World Politics* 54, no. 1: 99–118.
Kalyvas, Stathis. 2006. *The Logic of Violence in Civil War*. Cambridge: Cambridge University Press.
Kalyvas, Stathis, and Matthew Kocher. 2007. "How 'Free' is Free-Riding in Civil Wars? Violence, Insurgency, and the

Collective Action Problem." *World Politics* 59, no. 2: 177–216.

Kalyvas, Stathis, and Matthew Kocher. 2009. "The Dynamics of Violence in Vietnam: An Analysis of the Hamlet Evaluation System." *Journal for Peace Research* 46, no. 3: 335–55.

Kant, Immanuel. [1781] 2007. *Critique of Pure Reason*. Translated by Max Mueller. London: Penguin.

Kasaba, Resat. 1988. *The Ottoman Empire and the World-Economy: The Nineteenth Century*. Albany: State University of New York Press.

Kehoe, Vincent. 2000. "The Works of Captain Thomas Simes, 1767 to 1782." *Journal of the Society for Army Historical Research* 78, no. 315: 214–17.

Kemp, Peter. 1957. *Mine Were of Trouble*. London: Cassell.

Kemp, Peter. 1958. *No Colours or Crest*. London: Cassell.

Kemp, Peter. 1961. *Alms for Oblivion*. London: Cassell.

Keohane, Robert. [1984] 2005. *After Hegemony*. Princeton: Princeton University Press.

Kilcullen, David. 2009. *The Accidental Guerrilla: Fighting Small Wars in the Midst of a Big One*. New York: Oxford University Press.

King, Gary, Robert Keohane, and Sidney Verba. 1994. *Designing Social Inquiry: Scientific Inference in Qualitative Research*. Princeton: Princeton University Press.

Kinross, Lord. 1977. *The Ottoman Centuries: The Rise and Fall of the Turkish Empire*. New York: Morrow Quill Paperbacks.

Kinzer, Steven. 1985. "Organizing the Revolution." *New York Times*, June 30, 1985, 17.

Kinzer, Steven. 2008. *All the Shah's Men: An American Coup and the Roots of Middle East Terror*. Hoboken: John Wiley and Sons.

Kipling, Rudyard. 1886. *Departmental Ditties and Other Verses*. London: W. Thacker and Company.

Kissinger, Henry. 2015. *World Order*. New York: Penguin.

Knapp, Michael, Anja Flach, and Ercan Ayboga. 2016. *Revolution in Rojava: Democratic Autonomy and Women's Liberation in Syrian Kurdistan*. Translated by Janet Biehl. London: Pluto Press.

Knetsch, Joe. 2003. *Florida's Seminole Wars, 1817–1858*. London: Arcadia.

Kojève, Alexandre. [1957] 1980. "Capitalisme et socialisme: Marx est Dieu; Ford est son prophète." *Commentaire*, no. 9: 135.

Kojève, Alexandre. 1993. "Note sur Hegel et Heidegger." *Rue Descartes: Logiques de l'ethique*, no. 7: 35–46.

Krieg, Andreas, and Jean-Marc Rickli. 2019. *Surrogate Warfare: The Transformation of War in the Twenty-First Century*. Washington, D.C.: Georgetown University Press.

Kropotkin, Peter. [1892] 2015. *The Conquest of Bread*. Translated by Brighton. London: Penguin Classics.

Krulak, Charles. 1997. *Marine Corps Doctrinal Publication 1—Warfighting*. Washington, D.C.: Headquarters, U.S. Marine Corps.

Kuhn, Thomas. [1962] 2017. *Structure of Scientific Revolutions*. 4th ed. Chicago: University of Chicago Press.

Lamb, Christopher. 2018. *The Mayaguez Crisis, Mission Command, and Civil-Military Relations*. Washington, D.C.: Joint History Office.

Lambakis, Steven. "Reconsidering Asymmetric Warfare." *Joint Forces Quarterly* 36: 102–08.

Lambert, Andrew. 1990. *The Crimean War: British Grand Strategy Against Russia*. Manchester: Manchester University Press.

Larsen, Christopher, and Norman Wade. 2019. *OPFOR3-2 SmartBook—Red Team Army, 2nd Edition*. Lakeland, FL: The Lightning Press.

Lawrence, T.E. 1920. "The Evolution of a Revolt." *Army Quarterly and Defence Journal*: 1–23.

Lawrence, T.E. [1926] 1991. *Seven Pillars of Wisdom: A Triumph*. New York: Anchor.

Le Bon, Gustave. [1895] 2002. *Le psychologie des foules [The Crowd: A Study of the Popular Mind]*. New York: Dover.

Lenin, Vladimir. [1906] 1965. "Guerilla Warfare." In *Lenin: Collected Works*. Moscow: Progress Publishers.

Lenin, Vladimir. [1916] 2011. *Imperialism, the Highest Form of Capitalism*. Mansfield Centre, CT: Martino Publishing.

Lenin, Vladimir. 1918. *The State and Revolution*. Moscow: n.p.

LeRiche, Matthew, and Matthew Arnold. 2012. *South Sudan: From Revolution to Independence*. New York: Columbia University Press.

Levi, Margaret, and Barry Weingast. 2016. "Analytic Narratives, Case Studies, and Development." Paper presented at the World Bank-Princeton Workshop on Case Studies and Development Practice, Washington, D.C., February 2016.

Levitsky, Steven, and Lucan Way. 2010. *Competitive Authoritarianism: Hybrid Regimes after the Cold War*. New York: Cambridge University Press.

Levy, Jack, and William Thompson. 2010. *Causes of War*. Singapore: Wiley-Blackwell.

Liddell Hart, Basel. [1954] 1991. *Strategy*. New York: Meridian.

Liebermann, Oren. 2022. "How Ukraine Is Using Resistance Warfare Developed by the US to Fight Back against Russia." CNN. Retrieved from https://www.cnn.com/2022/08/27/politics/russia-ukraine-resistance-warfare/index.html.

Lind, William, and Peter Thiele. [1991] 2015. *Fourth Generation Warfare Handbook*. Kouvala, Finland: Castalia House.

Lind, William, Keith Nightingale, John Schmitt, Joseph Sutton, and Gary Wilson. 1989. "The Changing Face of War: Into the Fourth Generation." *Marine Corps Gazette*: 22–26.

Lipset, S. Martin. 1959. "Some Social Requisites of Democracy: Economic Development and Political Legitimacy." *The American Political Science Review* 53, no. 1: 69–105.

Loadenthal, Michael. 2016. "Interpreting Insurrectionary Corpora: Qualitative-Quantitative Analysis of Clandestine Communiques." *Journal for the Study of Radicalism* 10, no. 2: 79–100.

Locke, John. [1690] 1988. *Two Treatises on Government*. Edited by Peter Laslett. Cambridge: Cambridge University Press.

Lukens, Mark. 2010. *Strategic Analysis of Irregular Warfare*. Carlisle, PA: U.S. Army War College Strategy Research Project.

Luttwak, Edward. [1968] 2016. *Coup D'état*. New York: Penguin.

Luttwak, Edward. 1976. *The Grand Strategy of the Roman Empire*. Baltimore: Johns Hopkins University Press.

Luttwak, Edward. 1987. *Strategy: The Logic of War and Peace*. Cambridge: Harvard University Press.

Lynn, John. 1996. "The Evolution of Army Style in the Modern West, 800–2000." *The International History Review* 18, no. 3: 505–45.

Lynn, John, ed. 1993. *Feeding Mars: Logistics in Western Warfare from the Middle Ages to the Present*. Boulder: Westview Press.

Machiavelli, Niccolò. [1531] 2008. *Discourses on the First Ten Books of Livy*. Edited and translated by Julia Bondanella and Peter Bondanella. Cambridge: Cambridge University Press.

Magioncalda, William. "A Modern Insurgency: India's Evolving Naxalite Problem." *South Asia Monitor* 8, no. 140: 1–4.

Mahan, Alfred Thayer. [1890] 2019. *The Influence of Seapower Upon History, 1660–1783*. Middletown, DE: Pantianos Classics.

Mahdi, Muhsin. 1987. "Alfarabi circa 870–950." In *History of Political Philosophy*, third edition, edited by Leo Strauss and Joseph Cropsey, 206–27. Chicago: University of Chicago Press.

Mahoney, James. 2010. "After KKV: The New Methodology of Qualitative Research." *World Politics* 62, no. 1: 120–47.

Malicdem, Ervin. 2017. "Aftermath of the Battle of Marawi." Schadowl Expeditions. Retrieved from https://www.slexpeditions.com/2017/11/223-marawi-battle-structures.html.

Mallison, W. Thomas. 1983. "Aggression or Self-Defense in Lebanon in 1982?" *Proceedings of the Annual Meeting (American Society of International Law)* 77: 174–89.

Mansfield, Harvey. 1987. "Edmund Burke 1729–1797." In *History of Political Philosophy*, third edition, edited by Leo Strauss and Joseph Cropsey, 687–709. Chicago: University of Chicago Press.

Mao Zedong. [1937] 2005. *On Guerrilla Warfare*. Translated by Samuel Griffith. New York: Dover.

Marighella, Carlos. 1969. *Minimanual do Guerrilheiro Urbano [Minimanual of the Urban Guerrilla]*. Sao Paolo: n.p.

Martin, JoAnn, and Carolyn Nordstrom, eds. 1992. *The Paths to Domination, Resistance, and Terror*. Berkley: University of California Press.

Martowych, Oleh. 1950. *The Ukrainian*

Insurgent Army. München: Ukrainian Information Office.

Marx, Karl. 1844. *A Contribution to the Critique of Hegel's Philosophy of Right.* Paris: Deutsch-Französische Jahrbücher.

Matthews, Matt. 2008. "We Were Caught Unprepared: The 2006 Hezbollah-Israeli War." *The Long War Series Occasional Paper,* no. 26.

McChrystal, Stanley. 2013. *My Share of the Task: A Memoir.* New York: Penguin Portfolio.

McGuire, Randall. 1982. "Breaking Down Cultural Complexity: Inequality and Heterogeneity." In *Advances in Archaeological Method and Theory, Volume 6,* edited by Michael Schiffer, 91–142. New York: Academic Press.

McNeill, William. 1982. *The Pursuit of Power.* Chicago: University of Chicago Press.

Mearsheimer, John. 2001. *The Tragedy of Great Power Politics.* New York: W.W. Norton.

Mearsheimer, John, and Stephen Walt. 2007. *The Israel Lobby and U.S. Foreign Policy.* New York: Farrar, Straus, and Giroux.

Meiselas, Susan. 1997. *Kurdistan: In the Shadow of History.* Chicago: University of Chicago Press.

Melson, Charles, ed. 2019. *The German Army Guerrilla Warfare Pocket Manual 1939–45.* Oxford: Casemate.

Metz, Steven. 2017. "Abandoning Counterinsurgency: Toward a More Efficient Antiterrorism Strategy." *Journal of Strategic Security* 10, no. 4: 64–77.

Michaels, Jeffrey. 2011. "Dysfunctional Doctrines? Eisenhower, Carter and the U.S. Military Intervention in the Middle East." *Political Science Quarterly* 126, no. 3: 465–92.

Milgram, Stanley. 1969. *Obedience to Authority.* New York: Harper & Row.

Miller, Joyce. 1977. "The Syria Revolt of 1925." *International Journal of Middle East Studies* 8, no. 4: 545–63.

Moore, Barrington. [1967] 1993. *Social Origins of Dictatorship and Democracy: Lord and Peasant in the Making of the Modern World.* Boston: Beacon Press.

More, Thomas. [1516] 2005. *Utopia.* Translated by Ralph Robinson and edited by Wayne Rebhorn. New York: Barnes and Noble Books.

Morgenthau, Hans. 1948. *Politics Among Nations: The Struggle for Power and Peace.* New York: Knopf.

Mumford, Andrew, and Bruno Reis, eds. 2013. *The Theory and Practice of Irregular Warfare: Warrior Scholarship in Counter-insurgency.* New York: Routledge.

Murray, Williamson, MacGregor Knox, and Alvin Bernstein, eds. 1996. *The Making of Strategy.* New York: Cambridge University Press.

Musa, Ahmed and Cindy Horst. 2019. "State Formation and Economic Development in Post-War Somaliland: The Impact of the Private Sector in an Unrecognized State." *Conflict, Security, and Development* 19, no. 1: 35–53.

Nagl, John. 2005. *Learning to Eat Soup with a Knife.* Chicago: University of Chicago Press.

Nerguizian, Aram. 2017. *The Lebanese Armed Forces, Hezbollah and the Race to Defeat ISIS.* Washington, D.C.: Center for Strategic and International Studies.

Nietzsche, Friedrich. 1886. *Jenseits von Gut und Böse.Vorspiel einer Philosophie der Zukunft [Beyond Good and Evil: Prelude to a Philosophy of the Future].* Translated by Vincent Lombardo. Leipzig: Druck und Verlag von C.G. Naumann.

Nietzsche, Friedrich. [1911] 1999. *Also sprach Zarathustra [Thus Spake Zarathustra].* Translated by Thomas Common. Edited by Joslyn Pine. Mineola, NY: Dover.

North, Douglass, John Wallis, and Barry Weingast. 2013. *Violence and Social Orders: A Conceptual Framework for Interpreting Recorded Human History.* Cambridge: Cambridge University Press.

Öcalan, Abdullah, 2017. *The Political Thought of Abdullah Öcalan: Kurdistan, Women's Revolution and Democratic Confederalism.* London: Pluto Press.

O'Hanlon, Michael. 2000. *Technological Change and the Future of Warfare.* Washington, D.C.: Brookings Institution Press.

O'Hanlon, Michael. 2019. *The Senkaku Paradox: Risking Great Power War over Small Stakes.* Washington, D.C.: Brookings Institution Press.

Olson, Mancur. 1959. *The Logic of Collective Action: Public Goods and the Theory of Groups.* Cambridge: Harvard University Press.

Ordobadi, Mammad. [1911] 2011. *Years of Blood: A History of the Armenian-Muslim Clashes in the Caucasus, 1905-1906*. Beirut: Ithaca Press.

Orwell, George. 1938. *Homage to Catalonia*. London: Secker and Warburg.

Osanka, Franklin, ed. 1962. *Modern Guerrilla Warfare*. New York: Macmillan.

Ostovar, Afshon. 2018. *Vanguard of the Imam: Religion, Politics, and Iran's Revolutionary Guards*. Oxford: Oxford University Press.

Othen, Christopher. 2015. *Katanga 1960-63: Mercenaries, Spies, and the African Nation that Waged War on the World*. Stroud: History Press.

Palmer, Bill. 2018. *The Languages and Linguistics of the New Guinea Area*. Berlin: De Gruyter Mouton.

Parry, Jonathan. 2015. "Just War Theory, Legitimate Authority, and Irregular Belligerency." *Philosophia* 43: 175-96.

Parsa, Misagh. 2000. *States, Ideologies, and Social Revolutions: A Comparative Analysis of Iran, Nicaragua, and the Philippines*. Cambridge: Cambridge University Press.

Peritz, Aki, and Eric Rosenbach. 2012. *Find, Fix, Finish: Inside the Counterterrorism Campaigns that Killed bin Laden and Devastated Al Qaeda*. New York: Perseus.

Perloff, Richard. 2014. *The Dynamics of Persuasion*. New York: Routledge.

Pew Research Center. 2015. "Religious Composition by Country, 2010-2050." Pew Research Center. Retrieved from https://web.archive.org/web/20200615053333/https://www.pewforum.org/2015/04/02/religious-projection-table/2010/number/all/.

Pittaway, Mark. 2004. *Eastern Europe 1939-2000*. New York: Oxford University Press.

Plato. 1997. *Complete Works*. Edited by John Cooper and D.S. Hutchinson. Cambridge: Hackett.

Pollack, Kenneth. 2019. *Armies of Sand: The Past, Present, and Future of Arab Military Effectiveness*. New York: Oxford University Press.

Porch, Douglas. [1982] 2005. *The Conquest of Morocco*. New York: Ferrar, Strauss, and Giroux.

Publius. 1788. *The Federalist: A Collection of Essays Written in Favour of the Constitution*. New York: J. and A. McLean.

Qi, Duo, Jiaqiang Zhang, Xiaolong Liang, Zhe Li, Jialiang Zuo, and Pengfei Lei. 2021. "Autonomous Reconnaissance and Attack Test of UAV Swarm Based on Mosaic Warfare Thought." Paper presented at the 6th International Conference on Robotics and Automation Engineering (ICRAE), Guangzhou, China, 19-22 November 2021.

Qiao Liang and Wang Xiangsui. 1999. *Unrestricted Warfare*. Nt. Brattleboro, VT: Echo Point Books.

Quraishi, Humra. 2004. *Kashmir: The Untold Story*. New Delhi: Penguin India.

Rabinovich, Itimar, and Carmit Valensi. 2021. *Syrian Requiem: The Civil War and Its Aftermath*. Princeton: Princeton University Press.

Radpey, Loqman. 2021. "Kurdistan on the Sevres Centenary: How a Distinct People Became the World's Largest Stateless Nation." *Nationalities Papers*: 1-30.

Rajagopalan, Rajesh. 2007. "Fighting Fourth Generation Wars: The Indian Experience." In *Global Insurgency and the Future of Armed Conflict: Debating Fourth-Generation Warfare*, edited by Terry Terriff, Aaron Karp, and Regina Karp. London: Routledge.

Ramsey, Robert. 2007. *A Masterpiece of Counterguerrilla Warfare: BG J. Franklin Bell in the Philippines, 1902-1902*. Fort Leavenworth: Combat Studies Institute Press.

Rawls, John. 1971. *A Theory of Justice*. Cambridge: Harvard University Press.

Reid, Anthony. 2005. *An Indonesia Frontier: Acehnese and Other Histories of Sumatra*. Singapore: Singapore University Press.

Reid, Brian. 2006. *The Civil War and the Wars of the 19th Century*. New York: Smithsonian Books.

Reynolds, E. Bruce. 2005. *Thailand's Secret War: OSS, SOE, and the Free Thai Underground During World War II*. Cambridge: Cambridge University Press.

Rid, Thomas. 2010. "The Nineteenth Century Origins of Counterinsurgency Doctrine." *The Journal of Strategic Studies* 33, no. 5: 727-58.

Robinson, Paul. 2010. "Soviet Hearts-and-Minds Operations in Afghanistan." *The Historian*.

Rodriguez, Ana. 2009. *Dividing the Isthmus: Central American Transnational*

Histories, Literatures, and Cultures. Austin: University of Texas Press.

Rogers, Joshua. 2016. *Civil War and State Formation: Exploring Linkages and Potential Causality*. Bern: Swiss Peace Foundation.

Rogers, Paul. 2016. *Irregular War: The New Threat from the Margins*. London: I.B. Tauris.

Rosecrance, Richard, and Arthur Stein. 1993. "Beyond Realism: The Study of Grand Strategy." In *Domestic Bases of Grand Strategy*, edited by Richard Rosecrance and Arthur Stein, 3–21. Ithaca: Cornell University Press.

Rousseau, Jean-Jacques. [1755] 1999. *Discourse on Inequality*. Translated by Franklin Philip. Edited by Patrick Coleman. Oxford: Oxford University Press.

Rousseau, Jean-Jacques. [1762] 1968. *The Social Contract*. Translated by Maurice Cranston. New York: Penguin.

Said, Edward. 1978. *Orientalism*. New York: Vintage.

Salibi, Kamal. 1988. *A House of Many Mansions: The History of Lebanon Revisited*. Berkeley: University of California Press.

Savage, Jesse, and Jonathan Caverley. 2017. "When Human Capital Threatens the Capitol: Foreign Aid in the Form of Military Training and Coups." *Journal of Peace Research* 54, no. 4: 542–57.

Savell, Stephanie. 2019. *Where We Fight: US Counterterror War Locations 2017–2018*. Providence: Brown University Costs of War Project.

Schechla, Joseph. 1984. "Palestinian Chronology, March 1 to May 15, 1984." *Journal of Palestine Studies* 13, no. 4: 222–44.

Schelling, Thomas. 1960. *The Strategy of Conflict*. Cambridge: Harvard University Press.

Schelling, Thomas. 1978. *Macromotives and Microbehavior*. New York: W.W. Norton.

Schmitt, Carl. [1932] 1963. *The Theory of the Partisan*. Translated by A.C. Goodson. Lansing: Michigan State University Press.

Schubert, Frank. 2013. *Other Than War: The American Military Experience and Operations in the Post-Cold War Decade*. Washington, D.C.: Joint History Office.

Schumpeter, Joseph. [1919] 1951. *Imperialism and Social Classes*. Oxford: Oxford University Press.

Schumpeter, Joseph. [1942] 2008. *Capitalism, Socialism, and Democracy*. New York: HarperCollins.

Scott, James. 1985. *Weapons of the Weak: Everyday Forms of Peasant Resistance*. New Haven: Yale University Press.

Scott, James. 1999. *Seeing Like a State: How Certain Schemes to Improve the Human Condition Have Failed*. New Haven: Yale University Press.

Scott, James. 2017. *Against the Grain: A Deep History of the Earliest States*. New Haven: Yale University Press.

Senate Select Committee on Intelligence. 2014. *Committee Study of the CIA's Detention and Interrogation Program—Findings and Conclusions—Executive Summary*. Washington, D.C.: United States Senate.

Serge, Victor. [1951] 2012. *Memoirs of a Revolutionary*. Translated by Peter Sedgwick and George Paizis. New York: New York Review Books.

Seymour, Lee, Kristin Bakke, and Kathleen Gallagher-Cunningham. 2016. "E Pluribus Unum, Ex Uno Plures: Competition, Violence, and Fragmentation in Ethnopolitical Movements." *Journal of Peace Research* 53, no. 1: 3–18.

Shahak, Israel. 1997. *Open Secrets: Israeli Nuclear and Foreign Policies*. London: Pluto.

Shariati, Ali. 1977. *Religion versus Religion*. Translated and edited by Laleh Bakhtiari. Chicago: Abjad.

Shariati, Ali. 1979. *Schools of Thought and Action*. Translated and edited by Laleh Bakhtiari. Chicago: Abjad.

Shehadeh, Raja. 1988. "Occupier's Law and the Uprising." *Journal of Palestine Studies* 17, no. 3: 24–37.

Sherman, Jason. 2009. "Army Eyes Irregular Warfare Lessons from Israeli Gaza Operation." *Inside the Pentagon* 25, no. 15: 19–20.

Shinoda, Hideaki. 2018. "Peace-building and State-building from the Perspective of the Historical Development of International Society." *International Relations of the Asia-Pacific* 18: 25–43.

Showalter, Dennis. 2004. "The Prussian Military State." In *Early Modern Military History, 1450–1815*, edited by Geoff

Mortimer, 118–34. New York: Palgrave Macmillan.

Shurkin, Michael, Raphael Cohen, and Arthur Chan. 2022. *French Army Approaches to Networked Warfare*. Santa Monica: RAND.

Simatupang. 1972. *Report from Banaran: Experiences during the People's War*. Translated by Benedict R. O'G Anderson. Singapore: Cornell Southeast Asia Program.

Skocpol, Theda. 1979. *States and Social Revolutions: A Comparative Analysis of France, Russia, and China*. Cambridge: Cambridge University Press.

Sloan, Geoffrey. 2019. "The Royal Navy and Organizational Learning." *Naval War College Review* 72, no. 4: 124–49.

Sluka, Jeffrey. 2000. *Death Squad: The Anthropology of State Terror*. Philadelphia: University of Pennsylvania Press.

Smele, Jonathan. 2017. *The "Russian" Civil Wars, 1916–1926: Ten Years that Shook the World*. New York: Oxford University Press.

Smith, George. 2021. "A Message from the Deputy Commandant, Plans, Policies, and Operations." *Marine Corps Gazette*: 6–7.

Solahudin. 2013. *The Roots of Terrorism in Indonesia*. Translated by Dave McRae. Ithaca: Cornell University Press.

Spears, Russell, Tom Postmes, Martin Lea, and Susan Watt. 2001. "A SIDE View of Social Influence." In *Social Influence: Direct and Indirect Processes*, edited by J.P. Forgas and K.D. Williams, 331–50. Philadelphia: Routledge.

Speier, Hans. 1941. "The Social Types of War." *American Journal of Sociology* 46, no. 4: 445–54.

Stalin, Joseph. [1924/1965] 2010. *Foundations of Leninism: Lectures Delivered at the Sverdlov University*. Beijing: Red Star.

Stalin, Joseph. 1938. *Dialectical and Historical Materialism*. New York: Prism Key Press.

Staniland, Paul. 2015. "Militia, Ideology, and the State." *The Journal of Conflict Resolution* 59, no. 5: 770–93.

Stern, Jessica, and Megan McBride. 2013. *Terrorism after the 2003 Invasion of Iraq*. Providence: Brown University Costs of War Project.

Stoker, Donald, and Craig Whiteside. 2020. "Blurred Lines: Gray-Zone Conflict and Hybrid War-Two Failures of American Strategic Thinking." *Naval War College Review* 73, no. 1: 12–48.

Strauss, Leo. 1987. "Niccolò Machiavelli 1469–1527." In *History of Political Philosophy*, third edition, edited by Leo Strauss and Joseph Cropsey, 296–317. Chicago: University of Chicago Press.

Striffler, Steve, and Mark Moberg, eds. 2003. *Banana Wars: Power, Production, and History in the Americas*. Durham: Duke University Press.

Suhrke, Astri. 2008. "Democratizing a Dependent State: The Case of Afghanistan." *Democratization* 15, no. 3: 630–48.

Suhrke, Astri, and Antonio De Lauri. 2019. *The CIA's "Army": A Threat to Human Rights and an Obstacle to Peace in Afghanistan*. Providence: Brown University Costs of War Project.

Svolik, Milan. 2011. *The Politics of Authoritarian Rule*. New York: Cambridge University Press.

Tabatabai, Ariana. 2020. *No Conquest, No Defeat: Iran's National Security Strategy*. Oxford: Oxford University Press.

Taber, Robert. [1965] 2002. *War of the Flea: The Classic Study of Guerrilla Warfare*. New York: Potomac Books.

Tainter, Joseph. 1988. *The Collapse of Complex Societies*. Cambridge: Cambridge University Press.

Taylor, Brian, and Roxana Botea. 2008. "Tilly Tally: War-Making and State-Making in the Contemporary Third World." *International Studies Review* 10, no. 1: 27–56.

Thayer, Charles. 1963. *Guerrilla*. New York: New American Library.

Thompkins, Paul. 2013. *Undergrounds in Insurgent, Revolutionary, and Resistance Warfare*. Edited by Robert Leonhard. Fort Bragg, NC: USASOC.

Thompkins, Paul, ed. 2012. *Casebook on Insurgency and Revolutionary Warfare: Volume II: 1962–2009*. Fort Bragg, NC: USASOC.

Thompkins, Paul, ed. 2013. *Casebook on Insurgency and Revolutionary Warfare: Volume I: 1933–1962*. Fort Bragg, NC: USASOC.

Thompson, Elizabeth. 2013. *Justice Interrupted*. Cambridge: Harvard University Press.

Tilly, Charles. 1990. *Coercion, Capital, and European States*. Oxford: Blackwell.

Tilly, Charles. 2003. *The Politics of Collective Violence*. New York: Cambridge University Press.

Tilly, Charles. 2005. "Regimes and Contention." In *The Handbook of Political Sociology: States, Civil Societies, and Globalization*, edited by Thomas Janoski, Robert Alford, Alexander Hicks, and Mildred Schwartz, 425–40. New York: Cambridge University Press.

Timmermans, Stefan, and Iddo Tavory. 2012. "Theory Construction in Qualitative Research: From Grounded Theory to Abductive Analysis." *Sociological Theory* 30, no. 3: 167–86.

Toliver, Raymond. 1997. *The Interrogator: The Story of Hanns-Joachim Scharff, Master Interrogator of the Luftwaffe*. Atglen, PA: Schiffer Military History.

Tolstoy, Leo. [1912] 2003. *Hadji Murad*. Translated by Aylmer Maude. New York: Random House.

Tompkins, Paul, ed. 2013. *Casebook on Insurgency and Revolutionary Warfare Volume I: 1933–1962, Revised Edition*. Fort Bragg, NC: USASOC.

Tourison, Sedgwick. 1991. *Talking with Victor Charlie: An Interrogator's Guide*. New York: Ivy Books.

Traboulsi, Fawwaz. 2012. *A History of Modern Lebanon*. London: Pluto Press.

Trinquier, Roger. [1964] 2006. *Modern Warfare: A French View of Counterinsurgency*. Translated by Daniel Lee. Westport, CT: Praeger Security International.

Trotsky, Leon. 1930. *History of the Russian Revolution*. Translated by Max Eastman. Ann Arbor: University of Michigan Press.

Tuckman, Bruce. 1965. "Developmental Sequences in Small Groups." *Psychological Bulletin* 63: 384–99.

Tuckman, Bruce, and M. Jensen. 1977. "Stages of Small Group Development Revisited." *Group and Organizational Studies* 2: 419–27.

Tuker, Francis. 1948. *The Pattern of War*. London: Cassell.

Turbiville, Graham. 2007. "Hunting Leadership Targets in Counterinsurgency and Counterterrorist Operations." *JSOU Report* 07-06: 1–114.

Turbiville, Graham. 2009. "Guerrilla Counterintelligence: Insurgent Approaches to Neutralizing Adversary Intelligence Operations." *JSOU Report* 09-1: 1–92.

Ucko, David, and Thomas Marks. 2022. *Crafting Strategy for Irregular Warfare*. Washington, D.C.: National Defense University Press.

U.K. Ministry of Defence. 2021. *Red Teaming Handbook*. 3rd ed. London: UK MOD.

U.K. War Office. 1939a. *The Art of Guerrilla Warfare*. London: General Service (Research).

U.K. War Office. 1939b. *The Partisan Leader's Handbook*. London: General Service (Research).

United Nations Secretary-General. 1982. "Report of the Secretary General on the United Nations Interim Force in Lebanon (for the period from 11 December 1981 to 3 June 1982)." UN S/15194, dated 10 June 1982. United Nations. Retrieved from https://web.archive.org/web/20081202004040/http://domino.un.org/unispal.nsf/9a798adbf322aff38525617b006d88d7/687deff0fe05590a85257019006e4036!OpenDocument.

U.S. Army Special Operations Command. 2015. *"Little Green Men": A Primer on Modern Russian Unconventional Warfare, Ukraine 2013–2014*. Fort Bragg, NC: USASOC.

Van den Berg, Axel, and Thomas Janoski. 2005. "Conflict Theories in Political Sociology." In *The Handbook of Political Sociology: States, Civil Societies, and Globalization*, edited by Thomas Janoski, Robert Alford, Alexander Hicks, and Mildred Schwartz, 72–95. New York: Cambridge University Press.

Vine, David, Cala Coffman, Katalina Khoury, Madison Lovasz, Helen Bush, Rachel Leduc, and Jennifer Walkup. 2020. *Creating Refugees: Displacement Caused by the United States' Post-9/11 Wars*. Providence: Brown University Costs of War Project.

Voelz, Glenn. 2014. "Is Military Science 'Scientific?'" *Joint Forces Quarterly* 75, no. 4: 84–90.

Von Clausewitz, Carl. [1832] 2004. *Vom Krieg [On War]*. Translated by J.J. Graham. New York: Barnes and Noble.

Von Dach, Hans. 1957. *Der totale Widerstand: Eine Kleinkriegsanleitung für*

Jedermann [Total Resistance: A Guerrilla Warfare Manual for Everyone]. Translated by Hans Lienhard. Wales: Panther Publications.

Von der Heydte, Friedrich A.F. [1972] 1986. *Der Moderne Kleinkrieg als Wehrpolitisches und Militärisches Phänomen [Modern Irregular Warfare as a Phenomenon of Military Policy]*. Translated by George Gregory. New York: New Benjamin Franklin House.

Von Steuben, Friedrich Wilhelm. 1779. *Regulations for Order and Discipline of the Troops of the United States*. Boston.

Wagner, R.H. 2007. *War and the State: The Theory of International Politics*. Ann Arbor: University of Michigan Press.

Walt, Stephen. 1996. *Revolution and War*. Ithaca: Cornell University Press.

Waltz, Kenneth. [1959] 2001. *Man, the State, and War: A Theoretical Analysis*. New York: Columbia University Press.

Waltz, Kenneth. [1979] 2010. *Theory of International Politics*. Long Grove, IL: Waveland Press.

Waltz, Kenneth. 1993. "The Emerging Structure of International Politics." *International Security* 18, no. 2: 44–79.

Waltzer, Michael. 1977. *Just and Unjust Wars: A Moral Argument with Historical Illustrations*. New York: Basic Books.

Ward, Steven. 2014. *Immortal: A Military History of Iran*. Washington, D.C.: Georgetown University Press.

Weber, Max. [1918] 1946. "Politics as a Vocation." In *From Max Weber: Essays in Sociology*, translated and edited by H.H. Gerth and C. Wright Mills, 77–128. New York: Oxford University Press.

Weigert, Gideon. 1963. "The Kurdish Republic of 1946 by William Eagleton." *Die Welt des Islams* 8, no. 4: 283–4.

West, John. 2008. *Fry the Brain: The Art of Urban Sniping and its Role in Modern Guerrilla Warfare*. Countryside, VA: SSI.

Westmoreland, William. 1976. *A Soldier Reports*. New York: Doubleday.

White, Jeffrey. 1999. "Irregular Warfare: A Different Kind of Threat." *American Intelligence Journal* 17, no. 1/2: 57–63.

White House. 2017. *National Security Strategy*. Washington, D.C.: White House.

Wittfogel, Karl. 1957. *Oriental Despotism*. New Haven: Yale University Press.

Yesiltas, Murat and Necdet Ozcelik. 2018. *When Strategy Collapses: The PKK's Urban Terrorist Campaign*. Istanbul: SETA Publications.

Zarrinkoub, Abdolhossein. [1957] 2017. *Two Centuries of Silence*. Translated by Paul Sprachman. Costa Mesa, CA: Mazda Publishers.

Zimbardo, Philip. 2007. *The Lucifer Effect: Understanding How Good People Turn Evil*. New York: Random House.

Zonis, Marvin. 1983. "Iran: A Theory of Revolution from Accounts of the Revolution." *World Politics* 35, no. 4: 586–606.

Index

Numbers in ***bold italics*** indicate pages with illustrations

Abane, Ramadane 166
Abbas, Ferhat 100, 157, 190
Abkhazia, Caucasus 137
Abu Ghraib (prison in Iraq) 167
active defense 41, 91
Afghanistan 2, 6, 15, 34, 61–3, 83, 96, 100, 105, 107–8, 117, 120, 141, 168, 173, 182, 196
Africa National Congress 164
Ahmad, Sheikh 25
al-Abadi, Haidar Saber 167
al-Awdeh, Ahmed 157
al-Awlaki, Anwar 109
al-Baghdadi, Abu Bakr 109
Albania 114, 157
al-Bashir, Omar 162
al-Farabi 24
Algeria 7, 55, 71, 95–6, 98–101, 133, 140, 147, 165, 167, 170, 190, 195
Algerian Civil War 10, 139–41
Algerian War of Independence 10, 50, 54–5, 68, 71, 99–101, 130, 165–7, 199
Algiers, Algeria 50, 56
al-Hawza (The Shi'a Seminary) 170
al-Manar (The Lighthouse) 171
ALN *see* National Liberation Army
al-Qaeda 63, 65, 72, 109, 168, 170–1
al-Qaeda in the Islamic Maghreb 72, 139–40
al-Shammari tribe 122
al-Zarqawi, Abu Musab 109, 174
al-Zawahiri, Ayman 109
Amal Movement 117, 146, 158, 191
American Civil War 56
American Revolution 56–7, 58, 116
American Revolutionary War 56–7, 92, 198
American University Beirut 159
An Phoblat 170

anarchy 45–6, 88, ***128***, 178
L'anarchy (France) 170
ancien régime 5, 145, 181
Andhra Pradesh, India 141, 143
Andorra 137
Angola 56, 183
anomie 20, 35–7
Ansar al-Din 139, 141
Anya-Nya 156, 158
AQIM *see* al-Qaeda in the Islamic Maghreb
Arab Revolt 103
Arab Spring 115, 157, 182
Argentina 55–6, 145
Aristotle 35, 43
Armée de libération nationale see National Liberation Army (Algeria)
Armenia 133, 142
Armenian-Azerbaijani conflict 133
Armenian Dashnak Party 133
Army Design Theory 64
Artsakh, Caucasus 137
'*asabiyyah* 19, 35–7, 171–2
Assam, India 111
assassination 70, 109, 129, 154
asymmetric warfare *see* fourth generation war
Ataturk, Kemal 46
Aufstragstaktik see mission command/ mission-type orders
Aurocania-Patagonia 138
Austria 86, 89, 92, 94, 115
Authorization for Use of Military Force 80
Azawad 137, 139–41
Azerbaijan 133, 142
Azerbaijan Province, Iran 161, 182
Aztecs 36
Azzam, Abdullah 168

243

Index

Baader-Meinhof Group *see* Red Army Faction
Baarle-Hertog 137, 143
Ba'ath Party (Iraq) 165
Ba'ath Party (Syria) 165
Badr Corps 117
Baghdad Pact 109–10
Baluchis 101
Banana Wars 58, 145, 166
Banda Aceh 66, 137, 148
Bandenbekämpfung 53, 93–5, 168
bandenkrieg 92
Barzani, Mullah Mustafa 25, 161
Bastiat, Frédéric 134
Batista, Fulgencio 148, 158, 170, 197
Battle of Algiers *154*, 166, 190
Battle of Austerlitz 89
Battle of Borodino 39, 173
Battle of Cannae 43
Battle of Fleurus 89
Battle of Jena-Auerstedt 40, 88–9
Battle of Marengo 89
Battle of Omdurman 102
Battle of Ratisbon 89
Battle of Tuttlingen 92
Battle of Ulm 89
Bavaria 115
bazaari community 140
Beirut, Lebanon 64, 116–7, 132, 158–60
Belgium 137
Bella, Ben Ahmed 157
Berlin Conference 96
Bernays, Edward 166
Bin Laden, Osama 109
Black and Tans 104
bled 56, 101
blitzkrieg 91
Boer War 103
Bolivia 56, 163
Bolotnikov 132, 169
Bolshevik Revolution 47, 50, 56, 58, *113*, 181
Bolsheviks *112-13*, 114
Boot, Max *see* neoconservatism
branches and sequels 90
Brazil 56, 129, 157, 168
Britain *see* United Kingdom
British Protectorate of Iraq 103
British Sierra Leone Selection Trust 122
Bulavin 132, 169
Byzantine Empire 40

Cabezas, Omar 168
cafecultura 123, 190
Callwell, Charles 95–6, 98
Calvinism 193
Cambodia 100

cameral science 65
Camp Bucca (Iraq) 167
Canada *113*, 116
carpet bombing 103, 105, 108, 161
Casamance 137
Casbah 50, 140, 190
Case White and Case Black 114
Caspian Sea 67
Castro, Fidel 157–9, 169
Caucasus 67, *113*, 133, 195
center of gravity 89
centra gravitatis see center of gravity
Central Intelligence Agency 44, 79, 104, 110, 120, 166, 183
Ceuta and Melilla 138, 143
Chad 192
Chadians 102
Charleston, South Carolina 116
chaussers 92–3
Chechens 63, 109
Chechnya 109, 119
Chetniks 50
Chiang-Kai Shek 137
Chile 170
China 9, 19, 40–1, 56, 91, 103, 116, 140, 157, 179, 192
Chindits 6
Chinese Communist Party 91
Chinese Revolution 47, 56, 58
Chuyen, Le Xuan 106, 175
CIA *see* Central Intelligence Agency
Civilian Irregular Defense Group *194*
Cold War 9–11, 42, 74, 117, 120, 178, 183
collectivism 151
Colombia 137
Combat Hunter program 93
commandos noir 96
Communism 46, 55, 58, 73–4, 109–10, 147, 157, 165, 168, 174
concentration camps 100, 103, 105
Congo, Democratic Republic of 34, 56, 122
Cossack insurgency (1606) 169
Cossack insurgency (1708) 156, 169
counterinsurgency 6, 12, 48, 53–5, 58, 66, 68, 71–2, 81–2, 94–101, 103–4, 108, 110–11, *112*, 196
counterterrorism 61, 68, 79, 81, 108, *112*
coup d'état 47, 49, 58, *112*, 117, 119, 127, 129, 140, 145, 162–3, 166, 181
covert action 78–9, 114–5, 120, 151, 161
Crimea 92, 120–1
Crimean War 92
Cuba 137, 148, 170, 197
Cuban Revolution 55–6, 137, 147–8, 158, 168, 197

Index

Cyprus 49, 55, 103–4, 133, 137
Cyprus Emergency 55

Dabiq (Da'esh) 171
Da'esh 57, 63, 65, 69, 72, 119, 121, 137–8, 140, 151, 167, 171, 192, 199
Damascus, Syria 99
Damavand Line 68
Darfur, Sudan 159
Da'wah (proselytizing) 169, 171
De Beers Diamond Protection Force 122
Declaration of Independence 32
defense in depth 89–91
de Gaulle, Charles 100
de Malberg, Raymond Carré 180
denied area operations 70, 95
Denmark 137
Derg (Ethiopia) 168
Dhofar Rebellion 68, 145
dialectic 6, 14, 39–40, 42, 45–6, 51, 85, 196
Dien Bien Phu 71, 96, 100
dirty war (*le sale guerre*) 100
Donbas 120
Drif, Zora 190
Druze 99, 115–6, 139, 147
Druze Revolt 99, 105
Dudayev, Dzhokhar 109
Durkheim, Émile 20, 22, 69
Dutch *see* Netherlands
Dutch Caribbean 138
Dutch East India Company 122

East Timor *see* Timor-Leste
Easter Rebellion 102
Egypt 119, 133
Ehrhardt, Arthur 94, 168
Eisenhower Doctrine 109
El Salvador 110, 163
El-Sisi, Abdel Fatah 119
Emmerich, Andreas 92–3
Engels, Friedrich 193
English Statute of Monopolies **144**
EOKA (Cyprus) 103
Eritrea 142
espionage 70, 107–8, 151
esprit de corps 37, 40
état de police (police state) 180
état légal (legal state) 180–1
Ethiopia 105, 142, 168
ethos 35
Eurocentrism 9, 12, 40, 66–7, 88
Evian Conference 130

Fanon, Franz 156, 170, 174
FARC 137
firebombing *see* carpet bombing

First Anglo-Dutch War 91
First World War *see* World War I
FLN *see* Front de libération nationale
Florence 66
FMLN *see* Frente Farabundo Martí para la Liberación Nacional
foco theory 55–6, 178
foreign internal defense 68, 82, 96, *112*, 119, 194
Foreign Military Sales 83
Former Republic of Yugoslavia *see* Yugoslavia
Fort Bragg, North Carolina 68
fourth generation war 73–6, 96, 192
France 9, 19, 37, 40, 45, 54, 62, 65, 67–8, 71–2, 81, 86, 88, 90, 95–6, 98–105, 107, 110, *113*, 114–5, 121, 130, 132, 138–9, 146–7, 151, *154*, 157, 164, 179, 190, 193, 195
Frankfurt School 22
French Guiana 138
French Revolution 22, 47, 51, 56, 58, 98, 114, 193
Frente Farabundo Martí para la Liberación Nacional (El Salvador) 163
Front de Combat Hebreau (Hebrew Combat Front) 170
Front de libération nationale (Algeria) 99–100, 140, 147, *154*, 164–7, 170, 190
Fujimora, Alberto 166
Fulda Gap 118

Galula, David 54–6, 96
game theory 27, 73
Gaza Strip (Palestine) 118, 155–6, 159, 199
Geneva Conference 71
Geneva Conventions 63
Georgia 121, 137
Germany 11, 19, 37, 90, 93–6, 103, 110, 114, 121, 137, 171, 193
gharbzadehgi 24, 195
Giap, Vo Nguyen 25, 53, 105, 157–8
Gibraltar 137
Gilan, Iran 170
Global War on Terror 42, 48, 61–2, 64, 74, 105–6, 108–9, 153
globalization 72, 87
gloire 42, 153
Glorious Revolution 56, 89
Grand Armée 39–40, 93, 98
grano del oro (golden grain) 190
gray zone conflict *see* fourth generation war
Greece 10, 24, 147, 164
group dynamics *see* '*asabiyyah*
G.S.(R) 95
Guatemala 110, 145, 166

Index

Guatemalan War 36
Guayana Esequiba 137
guerrilla warfare 12, 42–3, 52–3, 55, 65, 69, 71–2, 90, 93–5, 100, 102, 104, 114–5, 147, 163, 168–9, 177
guerrillas *see* guerrilla warfare
guerrilleros see guerrilla warfare
Guevara, Ernesto "Che" 55–6, 147, 157, 163, 168, 170, 174, 178
gunpowder empires 67
Gustavus Adolphus 90

hafrada (Israeli apartheid system) 192
The Hague Convention 64
Haiti 117
Hama massacre 74
Hama model 74
Hamas 109, 117–8, 155–7, 184
Hannibal 43
Hanover 93
Hapsburgs 92
harkis 96
hasbara 158, 166
hawzah (Shi'a seminary) 170–1
hearts and minds 98–9, 104
Hegel, Georg W.F. 24, 27, 40, 42, 46–7, 51, 174
Hemingway, Ernest 168
Hessians 92
hezbollahi 70, 171
Hezbollah (Lebanon) 12, 14–5, 64–5, 109, 111, 117, 121, 137, 147, 158–9, 164–5, 171, 184, 191
hisbah police 138
Ho Chi Minh 99, 157
Hobbes, Thomas 27, 31, 43, *152*
Holy Roman Emperor 92
Honduras 145
Hornbeam Line 68
Hoxha, Envar 157
Huks 7, 66, 141, 165, 199
hybrid warfare *see* fourth generation war

Ibn Khaldun 20, 35
identity *152*
imperium in imperio 147
İmralı Island prison 49
inclusion 151
India 75, 111, 122, 141, 145
Indochina 62, 68, 96, 99–101, 157, 195
Indonesia 14, 18, 49, 55, 66, 148, 169, 179
Indonesian National Military 165
Indonesian National Revolution *see* Indonesian War of Independence
Indonesian War of Independence 14, 25, 148–50, 159, 165, 169, 199

Industrial Revolution 22, *144*
Inspire (al-Qaeda) 171
insurgency 33–4, 45–8, 53–5, 58, 61, 64–7, 69, 71–2, 74, 81, 88, 95, 97, 101, 104–5, 110–11, *112*, 115–6, 121, 140, 147, 151–3, *154*, 155–7, 159, 162–3, 165, 168–72, 174–6, 187, *188*, 191–3, *194*, 195, 197
insurgent motivations *154*
International Military Education and Training 83
Iran 9, 18–9, 40, 46, 49–50, 68, 70, 90–1, 101, 109, 140–1, 158, 160–1, 165, 170, 179, 182, 195
Iran al-Thawra (Revolutionary Iran) 170
Iran-Iraq War 142, 165
Iranian Revolution 7, 22, 47, 49, 56, 58, 73, 117, 133, 142, 167, 176
Iraq 48, 57, 62–3, 70, 83, 89, 96, 100, 103, 105–9, 117–8, 122, 137–8, 140–1, 151, 153, 160–2, 167–9, 174, 182, 186, 195, 199
Iraq Study Group Report 108
Irish Republican Army 102, 104, 164, 168
ISIS *see* Da'esh
Islam 72–3, 142, 148, 159, 173
Islamic Emirate of Kunar 141
Islamic Resistance (Lebanon) 164
Islamic Revolutionary Guard Corps 101, 117, 160
Israel 15, 49, 75–6, 96, 98, 107, 109, 111, 117, 158–9, 164, 166, 183, 192–3
Israel-Gaza War (2008) 118
Israel-Hezbollah War (2006) 199
Italy 46, 66, 86, 94, 105, 116, 193

Jaeger Korps 93
Jaegerkommando 93
Jagdkampf 92
Jaish al-Mahdi see Mahdi Army (Iraq)
Jamat ud-Dawah (Pakistan) 164
Jangal (The Jungle) 170
Japan 56, 91, 105, 108, 116, 137, 157–8, 183, 193
Java 14, 18, 106, 149–50, 159, 169
Jefferson, Thomas 32
Jihad (magazine) 170
Jihad Council (Lebanon) 164
jihadists 65, 71–2, 107, 139, 141, 145, 168, 170, 174
Jomini, Antoine-Henri 38–9, 61, 80, 86, 88–9, 91, 98, *185*, 197
Jordan 34, 63, 174
Jung, Carl 24

Kabinettskrieg 39, 92
Kant, Immanuel 27, 180
Kashmir 111, 145, 199

Katanga, Congo 122
Kemp, Peter 103, 114, 116, 168
Khan, Mirza Kuchik 170
Khomeini, Ayatollah Ali 47, 142, 167
kill webs *see* fourth generation war
Kipling, Rudyard 173
Kleinkrieg see small wars
Korea 137
Korean War 94
Korybko, Andrei 121
Kosovo 49, 89, 137, 163, 183, 195, 199
Kosovo Liberation Army 163
Kriegsraison 118
Kropotkin, Peter 45–6, 48, 146, 174
Kunar Province, Afghanistan 141
Kuomintang 91, 137, 157
Kurdistan 160–3, 170, 191
Kurdistan Workers Party 152, 163
Kurds 7, 25, 49, 158, 160–2

Landsturm Edikt 39
Landwher 92
lawfare *see* fourth generation war
Lawrence, T.E. 96, 103
Lashkar-e Taiba (Pakistan) 164
Latvia **113**
League of Nations 138
Lebanese Civil War 117, 121, 139, 147, 155, 158, 164
Lebanese Communist Party 155
Lebanon 15, 64, 75, 82, 99, 109, 111, 116–8, 121–2, 132–3, 137, 146, 155, 158–60, 164–5, 170, 184, **188**, 191
Le Bon, Gustave 45–6
legitimacy capital 187, **188**
Lehi see Stern Gang
Le May, Curtis 105
Lenin, Vladimir I. 41, 47, 49–51, 58, 148, 174
Levi-Strauss, Claude 136
Leviathan 31
Liberia 155
Libya 105, 109, 117, 123, 146, 192
Liddell Hart, B.H. 40, 44, 190
Liechtenstein 137, 180
liminal sovereignty 130–2, 141, 182, **189**, 191
limited war theory 115
Lithuania **113**
Locke, John 29, 31–2, 127, 134, 184
logos 35
low-intensity conflict 62
Loyalty to the Resistance Bloc (Lebanese political party) 164
Luftwaffe 68
lura (village headman) 150
luring in deep 91

Machiavelli, Niccolò 39, 58, 66, 81, 87, 193
Madagascar 98
Maghreb 66, 139
Magna Carta 127
Magsaysay, Ramon 7, 141, 157
Mahabad Republic 7, 18, 25, 161, 170, 182, 198
Mahan, Alfred Thayer 41, 61, 88, 91
Mahan, Dennis Hart 88
Mahdi Army (Iraq) 170
Mahdists (Sudan) 102
Majles ash-Shura (Lebanon) 164
Malayan Emergency 53, 66, 103–4, 177
Malaysia 6, 103, 177
Mali 72, 101–2, 138–41, 143, 146
maneuver warfare 86, 88–9, 94
manipulus 90
Mao Zedong 25, 36, 41, 53, 55, 62, 65, 90–1, 131, 157, 168–9, 186, 192
Maoism 54
maquis 121, 151, 171, 190
Marawi 118–9
marjah 171
Marian reforms 90
Marighella, Carlos 56, 129, 157, 168, 178
Maronites 117
Marx, Karl 51–2
Marxism 7, 22, 168, 174
Marxism-Leninism 10, 42, 51, 55, 58, 147–8, 168, 170, 178, 186, 195
maskirovka see military deception
Mayaguez incident 192
Mayans 36
Mayotte 138
McChrystal, Stanley 100
Meiji Restoration 56
Mensheviks 174
Mesopotamian civilization 178
Mexican Revolution 156
Mexico 116, 179
MI5 104
Milgram, Stanley 22
milieu of oppression 187
military deception 70, 121, 169–70
Mining Union of Upper-Katanga 122
MIR *see* Revolutionary Left Movement
mission command/mission-type orders 94
Mitterrand, Francois 100
MLN-T *see* Tupamaros
MNLA *see* Movement for the Liberation of Azawad
modernization theory 7, 46–8
Mogador 138
Mohammad Shah Pahlavi 7, 50, 133, 140, 161

248 Index

Monroe Doctrine 123
Montevideo Convention 63, 179
More, Thomas 32
Morice Line 68, 99, 101
Morocco 7, 71, 96, 98, 133, 138, 198
Mosaddegh, Mohammad 170
mosaic warfare *see* fourth generation war
Moscow, Russia 169
Mosul, Iraq 106, 151
Mouvement national de libération de l'Azawad see Movement for the Liberation of Azawad
Mouvement pour l'unicité et le jihad en Afrique de l'Ouest see Movement for Oneness and Jihad in West Africa
Movement for Oneness and Jihad in West Africa 139
Movement for the Liberation of Azawad 139–40
Movimiento de Izquierda Revolucionaria see Revolutionary Left Movement
Mozambique 183
MRLA (Malaysia) 103
Mughal India 67
Mughniyeh, Imad 109, 160
El Mujahid (The Struggler) 170
Mujahidin (Afghanistan) 6, 61, 101
Mujahidin (Sudan) 122
MUJAO *see* Movement for Oneness and Jihad in West Africa
Multidimensional Integrated Stabilization Mission in Mali 140
muqawama 158
mushrikun (non-believers) 173
Mussolini, Benito 46

Nagorno-Karabakh 142, 199
nahda (renaissance) 132
Napoleon 39, 65, 86, 88–9, 93, 150
Napoleonic Wars 39, 41, 67, 86, 88, 90, 99, 179–80
Nasrallah, Hassan 158, 160
National Front (Lebanon) 170
National Liberation Army (Algeria) 164–5
National Liberation Front (France) 164
National Mission Force 82
NATO *see* North Atlantic Treaty Organization
Nazis 94, 100, 114, 116, 121, 151, 164–6, 168, 170
neoconservatism 66–7
neorealism *see* Realism
nepantla 36
Netherlands 14, 18, 25, 90, 106, 137, 149–50, 158, 165, 168, 193

New Guinea, the island of 190–1
Newmont Mining Corporation 122
Nicaragua 9, 110, 145, 168
Niger 82, 179
non-linear warfare *see* fourth generation war
North Africa 71, 96, 101, 139, 147, 169
North Korea 70, 90–1
North Vietnam 69, 71, 105–6, 148, 175
North Vietnamese Army 69
North Atlantic Treaty Organization 63, 70, 90, 123
Northern Cyprus, Turkish Republic of *see* Cyprus
Northern Ireland 102, 104, 147, 170
Northwest Africa 62, 65, 72
nuclear warfare 38, 43, 66, 70, 72–4, 120

Öcalan, Abdullah 49, **152**
offensive Realism *see* Realism
Office of Liberation Movements (Iran) 117
Office of Propagation of the Catholic Faith 169
Office of Strategic Services (United States) 157
Office of the Martyr Sadr 170
oil spot strategy 71, 98, 105, 140
Oman 68, 145
Operation Ajax 166
Operation Allied Force 89, 163
Operation Barbarossa 90
Operation Barkhane 62, 65
Operation Blue Bat 116–7
Operation Bright Star 162
Operation Cyclone 120
Operation Desert Shield/Storm 89, 192
Operation Litani 117
Operation Morthor 122
Operation Odyssey Dawn 123
Operation Peace for Galilee 158
Operation Rolling Thunder 110
Operation Serval 65, 72, 101–2, 140
Operation Switchback 194
operations, actions, and activities 95
Operations Yellow and Red 91
Organization of Petroleum Producing Countries 122
Orthodox Christianity 67, 133
Orwell, George 116, 168
Ottoman Empire 40, 66, 67, 98, 103, 115, 131–3, 141, 147, 160

Pakistan 75, 109, 120, 137, 141, 145, 164, 179
Palermo 66
Palestine 49, 55, 103, 118, 137, 155–6, 166, 170, 184, 199

Palestinian Authority 183
Palestinian Islamic Jihad 117
Palestinian Liberation Organization 158
Panama 145
pancasila 148
Paris Commune (1871) 193
partisans 12, 39, 49–50, 54, 71, 92–4, 114, 173
pathos 35, 108
Peace of Westphalia *see* Westphalian sovereignty
Peasant's War (1525) 193
pemuda (youth) 150
Peninsular War (Spain) 93, 98, 102, 173
People's Army of Vietnam 69
People's Liberation Army (China) 7
People's Liberation Front (Greece) 164
Persia *see* Iran
Persian Gulf 34, 192
Peru 110, 166
Petraeus, David 100
Philippines 6, 9, 66, 83, 118–9, 138, 141, 159, 165, 183, 199
PKK *see* Kurdistan Workers Party
Plato 24, 32, 174
Portugal 93, 98, 183
positional warfare *see* maneuver warfare
Potemkin Mutiny 181
Povstanets (The Insurgent) 170
Powell Doctrine 89
proletariat 51–2, 174
propaganda 14, 97–8, 158, 166, 168–70
prospect theory 131
Protestant Reformation 193
Provisional Irish Republican Army 164, 168, 170
Prussia 39–41, 65, 67, 86, 88, 92
psychological warfare 73, 100
Pugachev 132
Puntland 137

Qaddafi, Muammar 109
Qasim, Abd al-Karim 157
quadrillage 98, 100, 103–5
quasi states 137
Quds Force 117, 160
Qutb, Sayyid 22

Rantzau, Josias 92
rational choice theory 27, 57
ratissage 96, 98–9
Rawls, John 14, 21–2, 25–8, 31–2, 184
Realism (theory) 9, 42–3, 48, 50, 57, 84, 87–8
Rechtsstaat (constitutional state) 180–1
Red Army Faction 171

relative deprivation theory 46, 48
RENAMO (Mozambique) 183
revolution in military affairs 74–6
Revolutionary Left Movement (Chile) 170
Reza Shah Pahlavi 7, 46
Robespierre, Maximilian 114
Rojava 137, 161, 164
Roman Empire 43, 90
Rousseau, Jean-Jacque 22, 31, 49
Royal Irish Constabulary 104
Royal Prussian Army *see* Prussia
RUF (Liberia/Sierra Leone) 155–6
Rumiyah (Da'esh) 171
Russia 9, 15, 39, 41, 45, 49–50, 70, 74, 77, 90, 93, 96, 98, 105, 109, **112–3**, 116, 119–21, 132, 140, 146, 156, 161, 165, 168–9, 182
Russia Company 122
Russian Civil War **112–3**, 114
Russian Revolution (1905–6) 133, 181
Russo-Japanese War 133
Russo-Turkish War 133

sabotage 70, 95, 107–8, 151
Sadr, Musa 146
Safavid Persia 67
Sahrawi Arab Democratic Republic 7, 198
Said, Edward 24, 158
Salafism *see* jihadists
sale guerre see dirty war
San Marino 137
Sandinistas 168
Sartre, Jean-Paul 193
Sassanian Persia 25, 36, 131
saturation point theory 55
Saudi Arabia 182
Scharf, Hanns Joaquim 68
Schleswig, Duchy of 137
School of the Americas 110
Schwerpunkt see center of gravity
scientism 64
Second Punic War 43
Second World War *see* World War II
Security Assistance 83, 151
Security Cooperation 83, 96, 151
security dilemma *see* realism (theory)
security forces assistance 68, 83
"selectorate" theory 55
Seminole Native Americans 134
Seminole Wars 134, 146
Serbia 134, 192, 199
Seven Years War 86, 92–3
Shamir, Yitzhak 166
sharia law 138–9, 142, 148
Shariati, Ali 15, 22–5, 35, 46, 174
Shi'a Islam *see* Islam
Shi'a militia groups 70, 73

Index

shura 141, 163
Sierra Leone 122, 155
Simatupang 25, 106
Simes, William 93
Sinjar, Iraq 141, 147, 160
Sinn Fein 146–7, 164
Slipchenko, Vladimir 74
small wars 54, 94
Small Wars Manual 95–6, 104
smallpox charts 71, 99
Smith, Adam 144
Social-Democrat Party (Germany) 170
Soleimani, Qasem 101, 160
Somalia 117, 137
Somaliland 137, 191
Songhai 139
South Africa 56, 164
South Korea 183, 193
South Lebanon Army 117, 121, **188**
South Ossetia 137, 199
South Sudan 7, 18, 49, 156, 162, 177, 182–3, 195
South Vietnam 104–5, 110, 116, 148, 182–3, **194**
South Yemen 183
Southeast Asia 66, 96, 100, 104
sovereign dysfunction 1, 7–8, 13, 15–6, 18–21, 28–34, 45, 50–1, 85, 112, 116, 119, 127, **128**, 131–7, 140, 142–3, **144**, 145–7, 149, 153, 158, 163, 165, 173, 176, 179, 182, 185–7, 193, 195–7, 199
Soviet-Afghan War 101, 117, 119–20, 168, 182
Soviet Union *see* Russia
Der Sozial-Demokrat 170
Spain 93, 98, 102, 114–6, 137–8
Spanish Civil War 114, 116, 168
special activities 64, 151
Special Operations Executive (United Kingdom) 95, 103
special operations forces 70, 79–80, 82, 93–5, 101–2, 168, 194
Special Operations Task Force 82
special reconnaissance 70, 93
spectrum of conflict 96
Spetsnaz 101
spetzpropaganda 120
SPLA *see* Sudan People's Liberation Army
stability operations 95, **112**
Stalin 148, 169, 174
Stern Gang 166, 170
strategic hamlet 105
sub-conventional warfare 111
subversion 78, 95, 100, 108, 151
Sudan 7, 15, 34, 122, 139, 143, 156, 158–9, 162, 179, 195
Sudan People's Liberation Army 122, 156, 159, 162, 168
Sun Tzu 39, 81, 90
Sunni Islam *see* Islam
Sweden 90
Switzerland 168
Sykes-Picot Agreement 122
Syria 34, 63, 70, 74, 99, 103, 115, 117, 121–2, 131, 133, 137–9, 147, 153, 157, 160–1, 165, 169, 179, 182
Syrian Civil War 131, 182
Syrian revolt 99

tache d'huil see oil spot strategy
Taif Agreement 158, 164
Taiwan 137
Taliban 2, 137, 163
Tanzania 192
tanzimat reforms 132
tawhid 37
Tentara Nasional Indonesia see Indonesian National Military
territorio libre 137
terrorism 45, 69, 72, 79–82, 107–8, 120
Thailand 137, 157
Third Reich 100
Thirty Years War *see* Westphalian sovereignty
thymos 153, 174
Tikrit, Iraq 107
Timor-Leste 15, 49, 183
Title 10 U.S. Code 79–80, 82
Title 10 U.S. Code §127e 79
Title 50 U.S. Code 79, 82
Tito 12, 114
TNI *see* Indonesian National Military
Tokyo, Japan 108
Transnistria 137, 199
Treaty of Utrecht 137
Trinquier, Roger 71–2, 96, 100, 122, 195–6
Trotsky, Leon 10, 127, 130–1, 191
true belief 153
Tuaregs 101–2, 139–40
Tunisia 133
Tupamaros (Uruguay) 170, 178
Turkey, Republic of 49, 90, 109, 160, 163, 179, 182, 198
Turkish War of Independence 198
Turkistan Islamic Movement 63

Uganda 122, 192
Ukraine 12, 109, 114, 120–1, 147, 156, 165, 170
Ukraine Insurgent Army 58, 109, 147, 156, 165, 170, 177–8
Umkhonto we Sizwe (South Africa) 164

Index

ummah 142, 173
unconventional warfare 6, 63, 70, 74, 78, 80, 82, 94–5, 97, 102–4, 110, *112–3*, 115, 120–1, 151, 157, 168, 194
Union Sacrée 37
UNITA (Angola) 183
United Arab Emirates 182
United Fruit Company 58, 122, 166
United Kingdom 19, 43, 67, 81, 92–3, 95, 98, 102–3, 107, 109–10, *113*, 114–6, 122, 133, 147, 161, 166, 168, 178
United Nations 18, 63, 122, 138–40, 159, 190
United Nations Charter 147, 179
U.S. Air Force Special Operations Command 83
U.S. Army 68, 94–5
U.S. Army Command and Staff College 54, 94
U.S. Army Special Forces 82, 106
U.S. Army War College 119
U.S. Constitution 56
U.S. Department of Defense 83, 96
U.S. Department of State 83
U.S. Marine Corps 37, 41, 61, 71, 93, 95, 104
U.S. Marine Special Operations Command 82
U.S. National Security Strategy 76
U.S. Navy 91
U.S. Navy Special Warfare 82
U.S. Special Operations Command 68, 82
Uruguay 178
Utopia 31
Uzbekistan 180

van Creveld, Martin 39–40, 75, 80
Vatican City 137
Venice 66, 115
vertical envelopment 70, 101
Vienna Convention 63
Viet Minh 25, 62, 71, 116, 164, 190
Vietnam 25, 62, 64, 68, 71, 97, 100, 104, 116, 157, 164, 182, 196
Vietnam War 25, 53, 62, 68–9, 71, 93, 104–6, 115, 148, 175, 182, 190, *194*
Volkskrieg 34, 39, 90, 94
voluntarism 181
Volunteer Army, Russian *113*
von Bismarck, Otto 170
von Bülow, Karl Wilhelm 95

von Clausewitz, Carl 11, 19, 32, 34, 38–43, 61, 63–4, 75–6, 80, 88–9, 92, 97, 168, 173, *185*, 191–2, 199
von Dach, Hans 168
von der Heydte, Friedrich 11, 168
von Freytag, Wilhelm 93
von Glahn, Gerhardt 49
von Gneisenau, August Neidhardt 88
von Mercy, Frantz 92
von Moltke, Helmuthe the Elder 88
von Moltke, Helmuthe the Younger 92
von Scharnhorst, Gerhard 88
von Seeckt, Hans 94
von Stülpnagel, Joachim 90

Wahhabism 72
Waltz, Kenneth 9, 20, 24
War of Austrian Succession 115
War of the Spanish Succession 115
War on Terror *see* Global War on Terror
War Powers Resolution 79
Washington, George 92
Weber, Max 22, 130, 136, 179, 184
Weinberger Doctrine 89, 192
Wellesley, Arthur 93
Wellington, Duke of 93, 98
Weltanschauung 11, 23–4, 54, 56, 184
West Bank (Palestine) 199
West Germany *see* Germany
Western Sahara 137
Westmoreland, William 69, 71, 100, 105
Westoxification *see gharbzadehgi*
Westphalian sovereignty 5, 9, 11, 16, 28–30, 44, 57, 63, 66, 79, 82–4, 87, 90–2, *112*, 130, 134, *135*, 136–8, 141–3, 145, 173–4, 178–80, 183, *185*, 191, 195, 198
White Armies, South Sudan 177
White Russians *113*, 114, 116
Wittfogel, Karl 181
World War I 37, 75, 87, 90, *135*
World War II 9, 17, 40, 44, 58, 68, 74, 87, 91, 94, 96, 105, 108, 121, 138, 151, 157, 159, 164, 182, 199

Yazidis 103, 141, 147, 160
Yemen 66, 82, 117, 182–3, 198
yu chi chan 53–5, 91
Yugoslavia 49–50, 114, 198

Zarrinkoub, Abdolhosein 173

www.ingramcontent.com/pod-product-compliance
Ingram Content Group UK Ltd.
Pitfield, Milton Keynes, MK11 3LW, UK
UKHW041936140426
5217IPUK00014B/502